Craft of the Inland Waterways

Anthony Burton

AN IMPRINT OF PEN & SWORD BOOKS LTD.
YORKSHIRE – PHILADELPHIA

First published in Great Britain in 2023 by
Pen and Sword Transport
An imprint of
Pen & Sword Books Ltd.
Yorkshire - Philadelphia

Copyright © Anthony Burton, 2023

ISBN 978 1 39907 080 5

The right of Anthony Burton to be identified as author of this work has been asserted by him in accordance with the Copyright, Designs and Patents Act 1988.

A CIP catalogue record for this book is available from the British Library.

All rights reserved. No part of this book may be reproduced or transmitted in any form or by any means, electronic or mechanical including photocopying, recording or by any information storage and retrieval system, without permission from the Publisher in writing.

Typeset in 11/14 pt Palatino by SJmagic DESIGN SERVICES, India.

Printed and bound in India by Replika Press Pvt. Ltd.

Pen & Sword Books Ltd incorporates the imprints of Pen & Sword Books Archaeology, Atlas, Aviation, Battleground, Discovery, Family History, History, Maritime, Military, Naval, Politics, Railways, Select, Transport, True Crime, Fiction, Frontline Books, Leo Cooper, Praetorian Press, Seaforth Publishing, Wharncliffe and White Owl.

For a complete list of Pen & Sword titles please contact

PEN & SWORD BOOKS LIMITED
47 Church Street, Barnsley, South Yorkshire, S70 2AS, England
E-mail: enquiries@pen-and-sword.co.uk
Website: www.pen-and-sword.co.uk

or

PEN AND SWORD BOOKS
1950 Lawrence Rd, Havertown, PA 19083, USA
E-mail: Uspen-and-sword@casematepublishers.com
Website: www.penandswordbooks.com

Contents

	Introduction	6
Chapter One	The Humber Keel	8
Chapter Two	The Norfolk Wherry	19
Chapter Three	The Thames Barge and More	29
Chapter Four	The Canal Age	46
Chapter Five	Steam Power	72
Chapter Six	Motor Boats	94
Chapter Seven	Passenger Boats	108
Chapter Eight	Ferry Boats	121
Chapter Nine	Boating for Pleasure	132
Chapter Ten	Building the Boats	146
Chapter Eleven	The Modern World	159
	Museums and Preserved Craft	165
	Select Bibliography	169
	Acknowledgements	170
	Index	171

Introduction

In this book, 'inland waterways' is a term that is taken to mean rivers and canals that carry or carried commercial traffic. The craft have been limited to those whose working lives were concentrated on such waterways, though a few were able to and did make short coastal passages as well. This excludes, for example, the colliers that regularly plied between Newcastle and London but were unable to venture further inland. As explained in the first chapter, people have been using rivers for transport for many centuries, but I have concentrated on the vessels that developed over a period of time and then the design became more or less standardised and survived in vessels we can still see today, even if no longer trading.

Like many kids, I enjoyed messing about in boats, but my local river, the Nidd at Knaresborough, had nothing more exotic to offer than rowing boats, though I managed to do a little dinghy sailing with a friend on a nearby reservoir. Later in life, my wife and I spent some time on canoeing holidays, but it was after sinking on a journey down the Wye that a friend introduced us to the idea of canal holidays, which was to prove the start of a lifelong fascination with inland waterways. I was fortunate enough in the early 1980s to be asked to write and present an 8-part TV series for the BBC – *The Past Afloat* – which gave me the chance to sail and steam on a variety of preserved working boats, some of which will feature in the following pages. That gave me a new enthusiasm for working boats – which I have always found far more satisfying than the more familiar yachts and motor cruisers that are generally bought or hired for pleasure trips. Since the series ended, I have gone back to several of these vessels, always with the idea of being involved in the working life of the boat. So, I have crewed on Thames barges on match days, for example, and probably shovelled many tons of coal into the boiler of a Clyde puffer. It is the result of this first-hand experience and the opportunities it has given me to talk to people who worked on vessels in their trading days that encouraged me to write this book.

I would like to thank all those skippers who allowed me to join in the working life of their boats, with a special thanks to Nick Walker of the *VIC32* with whom I have spent many weeks chugging up and down the west coast of Scotland and along the Scottish canals. A special thanks, too, to my old friend Mike Lucas, who was the founder of Mikron theatre company, who for many years have toured England's canals in their converted Grand Union narrow boat, presenting shows that they wrote themselves. For one of these, *I'd Go Back Tomorrow*, Mike and members of the cast interviewed many men and women who had lived and worked on the canal and Mike was kind enough to let me use some of that material in this book. I have talked to many people about the working life of the boats, but any mistakes that might come to light are entirely my own.

CHAPTER ONE

The Humber Keel

No one really knows just how long Britain's rivers have been used for transport, but they certainly were in use long before any written records were made. Archaeologists have suggested that the only way the blue stones from Wales could have been taken to Stonehenge was by some form of raft or boat, round the coast and up the river Avon to the site near Salisbury. That, however, can only be speculation. We can also guess how human beings ever took to the water at all. It must have been obvious that things could be moved by water far more easily than they could over land. A huge log might be seen floating down river – a log that would have required an enormous effort to haul over rough ground. This was actually expressed in figures after experiments in the eighteenth century to find the maximum load that could be moved by a single horse. If you put the load on the horse's back, then around $1/8$ of a ton was the best it could manage. Harness it to a cart on a rough road – and in the eighteenth century, most roads were rough – then it could haul just five times as much. But attach the horse to a barge on a river and it could easily pull as much as 30 tons. Our ancestors thousands of years ago had no statistics, but they would have instinctively seen the advantages of travelling by water. Quite how the idea evolved into a working boat is more difficult to work out.

Sitting on a log is not exactly a stable way of getting around – falling off logs being famously easy. If, however, you lashed logs together to form a raft, you could sit more comfortably, but rafts are difficult to manage. However, it would be possible, even with nothing more than stone tools, to hollow out a log and sit in it, which would be much more stable and easier to handle. Another option would be to make some sort of float. Assyrian wall paintings show men on a river, clinging to inflated animal skins. The earliest evidence we have of real boats made in Britain came when two were discovered, preserved in the muddy banks of the Humber at North Ferriby in 1938. Excavation was held up by the outbreak of war the following year and was only completed in 1946. Further work in 1963 revealed a third boat. Attempts to lift the first boat from its oozing grave ended in disaster as it collapsed in the process, though all the fragments were carefully removed, numbered and reassembled in what must have been a giant jigsaw puzzle. It was decided not to attempt to lift boat number two as a whole, but instead it was cut up into manageable chunks and then removed. Boat number three also collapsed during the lift and had to be reassembled, with the help of more modern technology.

The third boat has been the most closely studied and recorded. It was 43½ feet long. This was far more than just a hollowed-out log but contained features that would be part of wooden ships down the ages. The timber stretching down the bottom of the boat from stem to stern, the keel, was built up from two solid sections of oak, slightly curved upwards at either end and joined by a scarf joint – one in which the ends to be united are carved so that they overlap, rather than meeting head on. From this, planks were added to build up the sides, joined together by strips of yew. The actual joints between the planks were not unlike a modern tongue and groove, made watertight by packing with moss and covering by slats. Cleats in the keel indicated that these held cross battens to strengthen the floor. Frustratingly, we have no means of knowing how many planks or strakes were built up to create the sides, nor how bow and stern were shaped. What we do know, however, is that this was a sophisticated craft, probably capable of making short sea voyages, and that it was constructed in the Bronze Age, somewhere about 1500 BCE.

To get a glimpse of what boats in Britain were like in the distant past we now have to move forward for over a millennium to the Iron Age. Several substantial log boats have been found from this time, but it is with the arrival of the Romans that we have evidence of vessels that seem in many ways much closer to the sailing barges of the modern era. The most interesting find, in many ways, was a vessel found in the Thames at Blackfriars in 1962. This one was incomplete but was estimated to have been 47ft long and 22ft beam. She was carvel built, that is the planks of the sides abutted each other, rather than overlapping, and were held to the frame by iron nails. There was a central cargo department, which was actually full of building stone, suggesting that the vessel had sunk. One third of the way from the stem was a slot to hold a mast. A coin found on board was dated to 88 CE. Experts have worked out that she was capable of carrying loads up to about 50 tonnes. So, we have what can reasonably be described as a sailing barge – and we know from other sources that Roman vessels were square rigged – a single sail suspended from a yard arm that ran at right angles to the central line from bow to stern. Some features would be unlike those of a modern vessel. Steering for example, was by a board set at one side of the vessel. These devices lasted right through the medieval period and were known as steer boards. To avoid damaging the steer board, ships always tied up with the opposite side of the ship next to the harbour wall – so there was a steer board (starboard now) side and a port side. However, the Roman craft had so much in common with the sailing barges we shall be looking at in this book, that it makes a useful starting point. We can start with returning to the Humber and a vessel that a Roman sailor would not have been too surprised to see, the Humber keel.

The keel is as close as we can get today to the ships of medieval England, and the name takes us back even further as it derives from the Anglo Saxon word for 'ship' – ceol. The first keels were wooden vessels and generally clinker built. That is to say that the planks overlapped, as in the North Ferriby Bronze Age boats. All keels are, by definition, square rigged with a single mast. As with the Roman cargo ship, this is set one third of the way back from the bows. The vessel we shall be concentrating on is *Comrade*, originally owned by Fred Schofield. In 1988 he wrote a very full account of all such vessels in his book *Humber Keels and Keelmen*. No one could be better qualified to write such a book, as he noted in the first chapter. 'My father Arthur Schofield's keel *Fanny* was the first that I embarked on, at the tender age of twenty-one days in 1906.' That keel had been launched in 1866, a clinker-built vessel, though changes were already being made in construction techniques. Clinker building was giving way to carvel construction and towards the end of the nineteenth century, the wooden knees that attached the beams to the side of the vessel were being replaced by iron forged by a blacksmith, and the keelson, the timber that stretched above the keel from stem to stern, was replaced by an iron girder. Later wooden hulls went out altogether to be replaced by iron and steel, but the basic hull shape remained unchanged.

Keels varied according to the waterways on which they worked. Until the middle of the sixteenth century, different river levels were overcome by flash locks, built into weirs. The water would build up behind the weir, but when a boat needed to pass, planks were removed from a gate set in the weir, allowing the water to rush down – the flash. Boats going downstream would run with the flash; those heading in the opposite direction would be winched up. The only constraint on boat size was the width of the opening. The first river to receive the modern type of lock, the pound lock, with a chamber and gates at each end, was the Lee in 1576. Over the years, flash locks were replaced on Britain's rivers and the eighteenth century saw navigation increased by the construction of artificial canals. Once a lock was built then that determined the maximum size of boat that could use that waterway. Fred Schofield's boat *Comrade* is known as a Sheffield boat, because it was designed to fit the locks on the waterways leading to the steel city. The limiting factor was the size of lock on the final stretch, the Sheffield & South Yorkshire Navigation, where the maximum length was 60ft 6in, width 15ft 6in and depth 6ft. To allow the maximum space for cargo, the keel has very bluff bows and rounded stern – you could think of it as rather like an oversized date box. The keels that were built for use on the Trent, for example, were quite different, although they originally had the same sail arrangement. They were about 74ft long and just over 14ft wide, sharply pointed at both bow and stern. Some had an extra mast in the stern – the mizzen – with a small lugsail. These were known as 'catches'. All keels were steered using a tiller, rather than a wheel. In this chapter, however, we shall be concentrating mainly on the Humber keel.

The distinguishing feature of the keel is the single mast, set just as it was in the Roman cargo barge, a third of the way from the bow. *Comrade* is square rigged with a large mainsail, suspended from a yard with a smaller topsail above that. A few keels had a top gallant above the topsail. The rigging is a mixture

The keel *Comrade* enjoying excellent sailing weather in the mouth of the Humber. The author is standing in the stern.

of standing and running rigging. The standing rigging is, as the name suggests, permanently in place, used to stabilise the mast. The ropes, which in the case of this vessel are actually made of wire, are tightened through dead-eyes. These are blocks with three holes, usually oval in shape, but on the keel are pear shaped. The running rigging is used to control the yards and sails. The yards are raised and lowered by halyards – an obvious derivation from haul-yards. The sails are controlled through the sheets, so that they can be left either at right angles to the hull or brought almost to a fore and aft line. Without this control, the vessel would be unable to function.

It seems to be a common misconception that square rigged ships rely on a following wind to blow them along. They are thought to be like the little toy boats children used to make with a matchbox, a match for the mast and a paper sail; blow on it and it would scut across the soapy bath water. It worked because if you wanted it to go in a different direction, you just had to blow a different way. Unfortunately, the wind cannot be controlled, so the only way a vessel can make progress when the wind is not blowing from astern is to tack, following a zig-zag course instead of a straight line. I remember Fred Schofield telling me with pride that his keel could sail as close to the wind as any other sailing vessel.

Controlling the gear requires winches, eight of them in all, together with an extra windlass for the anchor. To the Keelmen they were known as 'rollers'. Tacking is a complex business, and I can do no better than let Fred Schofield explain it in his own words. It may help to know that a vessel is on a starboard tack if the wind is coming from that side, and that the sheet pulls the sail and the yard in towards the stern, and the tack towards the bows:

> The captain would first shout to the mate to 'stand by' and would put his tiller 'hard down' to bring the ship up into the wind. When the sail starts to flutter, he would let go the lee sheet (say, the port sheet, assuming the ship has been on the starboard tack). The mate would then need to 'raise the tack' (that is, release the pawl on the starboard tack roller and let the tack run out). He would then take up some of the slack on

the port tack, bringing the port clew [the bottom corner of the sail] up to the shrouds. When the ship had passed through the eye of the wind, the captain would order the mate to 'let go the bowline'. This is an order which would always be given by the captain, and no mate, however experienced, would ever anticipate it. The reason is that the captain would have to haul the yard using the brace, and proper timing was vital. As soon as he had let go of the bowline the mate would need to 'get the tack down' (that is, the port tack) sharp and the captain would heave in the lee (starboard) sheet, 'sailing her full' for a short time to allow the ship to gather way again. The mate would 'pass the bowline over the forestay' and, when the ship had gathered way again, he would 'haul the bowline and make it fast'.

This may all seem a bit complex, but that is unavoidable, for it is a complicated manoeuvre. Basically, the crew are moving the sails, so that the corner of the sail that had been closest to the bows would end up nearer the stern and vice versa, and all this has to be done while keeping the ship sailing smoothly and without losing too much way. I have not seen this done by the captain and a professional mate, but I have watched the volunteers who work with Fred Schofield and have always been impressed by how efficiently everything was done. But Mr Schofield said to me that although he had taught them everything himself, 'They'll never sail as I did.' And, of course, he's right, if only because the professionals had to sail whatever the conditions in order to earn a living.

Moving the yards and sails is not the only work involved in tacking. Because the keels operate in the comparatively shallow waters of canals and not just in the deep of the Humber estuary, they have no keel at the bottom of the boat. So, to prevent them drifting sideways, they have two lee boards, one to each side of the vessel. These are heavy wooden devices, roughly pear shaped. When sailing, the lee board is dropped to the lee side of the vessel and the weather board raised. *Comrade* is an exception in that she has no lee boards. In the narrower stretches of the Stainforth & Keadby and Sheffield & South Yorkshire canals, tacking is not a practical option and sails can only be used with a favourable wind. In those circumstances, the keel has to be hauled along from the towpath. Horses were generally used, and the men who worked with them were known as the horse marines. But on some occasions, the crew would put on a special harness and pull the vessel along themselves.

There is a comfortable cabin below the after deck, accessed by a steep ladder. In the confined space, everything had to be neatly stowed away. Fred Schofield described one cabin as having fourteen cupboard doors and five drawers. A double bed was tucked away behind one set of doors. A cast iron stove was used for heating and cooking and at night an oil lamp suspended from an iron bar supplied the light. There was also a smaller cabin, the fo'c'sle, in the bows. Normally, the captain would be out all week, but was always home at weekends. Sometimes the family would come on board, and a simple bed could be made in the aft cabin. On those occasions, the mate was not needed, especially when only working on the canals. But when going to sea, a purchaseman was employed, who was paid by the day and provided with food when on board.

A pair of Humber keels on the Sheffield and South Yorkshire Navigation. The nearer keel can be seen to have a lee board raised. They are fortunate to have the wind in their favour, otherwise they would have had to be towed by horses.

Comrade in her working days when owned by Fred Schofield, taking on cargo at Hull.

Fred Schofield made a list of all the different cargoes carried in *Comrade* over the years, 75 different types from alum to wheat. The latter was one of the regular cargoes, the grain arriving in Hull from around the world. Loads of ninety tons were regularly carried to the mills around Sheffield Basin and to the Town Mills of Rotherham. Coal was an important part of the trade, travelling in the opposite direction from the collieries of South Yorkshire. One of the more unusual cargoes was basic slag from the steel industry, which was used as a fertiliser. Loading it was particularly difficult. This took place at Keadby, where the Stainforth and Keadby Canal joins the River Trent. The slag came by rail in hundredweight bags, which had to be moved by barrow from the trucks to the keel along a gangplank. As more and more bags were added, so the keel would sink lower in the water and had to be moved further from the bank to avoid grounding. By the time the final load had been added, the keel could have been as far as eighteen feet from the bank, and running a barrow for that distance up a steeply sloping plank must have been a difficult and dangerous activity. The one thing that is clear is that Keelmen never suffered from lack of variety in their work.

Comrade continued to trade right through the 1960s, but more and more inland waterways craft were being scrapped. There were, however, enthusiasts who were keen to preserve the traditional craft of the Humber. In 1974, ownership of the keel passed to the Humber Keel and Sloop Preservation Society, who went ahead with a major restoration programme to ensure that she continues to sail to this day. The 'sloop' in the Society's name is the *Amy Howson*. Basically, the hull is very similar to the keel's but with rather finer lines aft than the rounded stern of the keel. The rigging, however, was completely different. Roger Finch 'in his book *Sailing Craft of the British Isles*' wrote that 'the sloop was a much easier craft to handle in the tideway' than

Fred Schofield at the tiller of *Comrade*. He continued to skipper the vessel after it was taken over by the Humber Keel and Sloop Preservation Society and taught the volunteers how to sail her.

Comrade's sister ship, *Amy Howson*. She has the same basic hull as the former, but is a sloop, fore and aft rigged instead of square rigged like *Comrade*. She too is regularly sailed by the Preservation Trust.

the keel – a statement which I suspect would be challenged by Fred Schofield. One of the sloop's activities was dredging for sand, using a bag with the mouth held open by an iron hoop. This was suspended from a chain and dragged along the bottom of the river. They also worked on a gravel bed in the mouth of the Humber, going aground on the ebb tide, digging out the gravel, and sailing off again on the rising tide. Like *Comrade*, *Amy Howson* is regularly sailed by the Trust. Under sail, the resemblance to the keel is less marked, for she is fore and aft rigged, a form of rigging which will be seen in the next craft, the Norfolk wherry.

Chapter Two

The Norfolk Wherry

Rather confusingly, the word 'wherry' in many parts of Britain referred not to a sailing barge but to a rowing boat for passengers. On the Thames in London, for example, the name had a very specific meaning; it was a passenger boat rowed by more than one man, usually three or four. In the early seventeenth-century picture of Yarmouth (page 20) just such a passenger wherry can be seen, rowed by two men. The cargo vessels seen in the same picture are keels, though unlike the Humber keel, they have no topsails. The Norfolk and north Suffolk rivers were certainly used for transport by the Romans and probably in earlier times as well, but the landscape then would have been very different. Much of the area was heavily forested, but in medieval times, the forests were steadily diminished for fuel and building, and huge areas of peat were dug up for fuel. By the fourteenth century, peat digging had to stop as the workings became flooded to form a series of lakes, the Norfolk Broads. This hugely increased water traffic as the Broads could be used as connecting links between the rivers, so that the navigable waterways stretched for some 250 miles (400km) from the most southerly point at Bungay, west of Lowestoft, to Antigham Ponds near North Walsham. Norwich was the great centre for manufactured goods, but Great Yarmouth and Lowestoft were the main seaports. The system did not only deal with overseas trade, as every village or town on the waterway system would have had its own staithes where essentials could be loaded and unloaded. These were the lakes and river which, from the eighteenth century onwards, would be served by the sailing wherries.

It is not clear exactly when the Norfolk wherry developed, but it had certainly reached its final stage of development by the nineteenth century. Everything about the wherry was a contrast to the Humber keel. Instead of the square sides and

James Corbridge's *Prospect of Yarmouth* of 1725 shows simple keels out on the water at the junction of the North River and the Yare. They are unlike the Humber keels in having their single mast set amidship. In among them is a passenger wherry with four oarsmen.

rounded bow and stern, the wherry had altogether finer lines, with a more pointed bow, though some had square transom sterns. The mast was stepped close to the bows, and carried a single sail set fore and aft instead of square. Almost every hull was clinker built, usually with fourteen two-inch overlapping oak planks to each side, though a few larger vessels had sixteen. Paradoxically, the vessel we shall be concentrating on is one of the very rare exceptions. *Albion* was carvel built at Oulton Broad in 1898. There are just two beams in the hull, one at each end of the hold, supported by substantial knees, both above and from below. Aft of the hold is the small, cosy cabin with two bunks and a stove, with a removable chimney. There is another small compartment in the bows, usually used for storing items for the crew. The bow itself is traditionally painted with a white quadrant.

The Norfolk wherry *Albion* photographed from a dinghy being towed behind. Unlike the barge seen in the previous illustration, the wherry has a mast set well forward and fore and aft rigging. The black sail is traditional for wherries.

The hold occupying most of the middle section of the vessel has two bulkheads, one forward and one aft. The hatches are raised above the deck by coamings, with space in either side for a narrow stretch of decking. The edge of the deck was called the binns and protected by binn irons. Everything on a wherry seems to have a special name unknown on other vessels. The coamings, for example, are the standing right-ups and on top of these were the shifting right-ups that could be removed for loading. The hatch covers mostly interlock, but to remove them at least one had to be left free so it could be easily lifted out – the taking-up hatch. The hull on *Albion* is 48ft long, 15ft wide and has a comparatively low draught of 4ft 6in and the cargo capacity is 23 tons.

The single mast is set just forward of the hatches, secured in a structure called a tabernacle. The mast on *Albion* rises 40ft above the deck and carries an impressive sail which, when fully extended, has an area of 1,400 square feet, and like all wherry sails is coloured black. There are rows of tabs sewn into the sail, so that it can be shortened if necessary in high winds – rolled up and tied in position, reefed. The sail itself is carried on a single yard, the gaff, with a jaw at one end to fit round the mast, from which it rises at an angle. It might seem that with just a two-man crew, handling such an expanse of canvas would be difficult, but everything is very well organised.

A single winch and halyard are used for raising the gaff, controlled through two blocks. The one nearest the mast, the crutch block, is raised until it can go no further and the luff, the edge of the sail nearest the mast, is taut. Then the second block, further out, comes into play, raising the peak, the top corner of the sail. So, when raising the gaff, one crew member works the winch, while the other takes the gaff line that hangs from the peak, and which is used to stop the peak swinging about as it is raised. At the top of the mast is a vane and each wherry had a distinctive little wrought iron figure, in this case, a woman in a billowing dress, and a pennant, about 6ft long.

One feature of the mast is hidden from sight. The wherry often has to pass under low bridges, so the mast has to be lowered. At the foot of the mast is a counterweight, generally

Albion preparing to lower the mast. The author is standing on the hatches with the line controlling the descent of the spar. The mast will then be lowered to the hatches by means of the winch in the bows.

a great lump of lead. Roger Malster 'in his book *Wherries and Waterways*' recounts that on one wherry this weighed in at 1 ton 12 cwt (1,626 kg). This makes the whole procedure of lowering and raising the mast comparatively simple. The mast is always slightly lighter than the counterweight, so when pulling it down, gravity does most of the work and it is then secured along the top of the hatches. Raising it requires no effort at all; thanks to the counterweight it simply rises slowly up until stopped by the tabernacle and stands vertically again.

As in all sailing vessels, a following wind makes life easy for everyone on board, but when sailing river reaches this is a luxury that is seldom enjoyed. All it needs is a sharp bend and, although the wind remains the same, it is no longer coming from the stern. The long pennant at the masthead that had fluttered

A working wherry on a day without a favourable wind. The mast has been lowered and the two crew members are moving the vessel along by quanting, pushing against the river bed with their long quant poles.

forward, now streams out to one side. The wherry has to tack. The sheets are controlled through sheet blocks that slide along an iron bar, the 'horse' set athwart the vessel and right in front of the steerer, who stands in the well by the cabin door. On any tack, the block will be at one end of the horse and the sail stretched out over the water on that side. The wherry will be heading towards one bank, so it is imperative that tacking can be accomplished quickly. Here, the fore and aft rig comes into its own, as, unlike the keel, the wind can be taken on either side of the sail. As the tiller is put over, the sail flutters loose for a moment, then as the wind catches it, the block slides across the horse and the sail takes up position on the opposite side of the wherry. It can be helped by the mate using the quant pole to push against the bank to help bring the bows round. This pole is mainly there, however, for a very different purpose. It is used when the wind simply does not blow at all.

Albion has never been fitted with an engine, and the rivers and lakes of the Broads have no convenient towpath for hauling the vessel along in a dead calm. The answer is quanting. The quant is a long, usually 22ft, pole, with a rounded knob at one end and is forked at the other. Each crew member has a pole and takes up position in the bows on either side of the vessel. The poles are thrust down into the water to the river bed, the men nestle the butt ends into their shoulders and walk towards the stern. At the end of the walk, they twist the poles – the forked end stops it sticking in the mud – pull up the pole and return to the bows to repeat the operation. It takes a bit of getting used to. When I first tried, I found that if you didn't thrust the quant pole down firmly and vertically, it simply floated uselessly to the surface. Having mastered that, it seemed not too tiring and quite fun, but that soon became less entertaining as time went by. But I was quanting a virtually empty vessel. With a full cargo on board, the boat would not only have been heavier but lower in the water, increasing the water resistance. Malster has an account of a wherry man, Walter Powley, who had a load from Norwich that he collected on a Saturday afternoon that had to be at Great Yarmouth by Monday morning. They left around 5.30 p.m. with no wind for the thirty-mile journey, and quanted all the way to Seven Mile House – named because it was seven miles from Great Yarmouth. Then, which must have been a huge relief, they got a favourable wind, and arrived at the port at 11 o'clock on the Sunday morning. They were probably helped by working with an ebbing tide.

Although the wherries were available for any cargo that needed shifting, there were some that were particular to the

Albion being quanted. This photograph shows one of the crew with the pole pressed against his shoulder starting to walk back towards the stern, keeping the other end of the quant on the river bed.

area. One of these was reeds for thatching. The reeds that grow in profusion on the edges of rivers and broads were generally cut from special reed barges, but when carried over long distances were carried in wherries. The reeds once cut were tied into bundles and a wherry might carry as many as 2,000 of these. And, because this was a largely agricultural area, there was a brisk trade in items that were needed, some perfectly pleasant, such as straw for stables and sand, some rather less so, such as manure – and the artificial manure manufactured in Norwich was particularly noxious. The oddest cargo was ice. When the dykes that criss-crossed the area froze, together with the upper reaches of the river, the wherry men used dydles, the wire scoops normally used in the fens for scooping out mud, to break up the ice, and load into the holds. This would then be taken to Great Yarmouth or Lowestoft, where it was used for preserving fish.

Sailing on the wherry is an exhilarating experience, with a frisson of unpredictability. On one occasion, we set off in a flat calm and quanted, then we got a light breeze and raised sail and by the end of the afternoon we had reefed the sail right down and were bowling along in fine style. There is, it has to be said, a certain smugness in being aboard the most magnificent vessel in the whole of the Broads. But by the end of the nineteenth century, the idea began to develop that the wherry could be adapted not for cargo but for pleasure. It seems that the first pleasure cruise took place around 1860 when a wherry hold was cleaned up and fitted out as living accommodation for a three-week cruise, before being returned to trading. Permanently fitting out a wherry for pleasure boating came a little later. In 1888, Press Brothers of North Walsham were advertising five wherries for hire, with two cabins, one able to accommodate 4 to 6 gentlemen and the other 3 to 4 ladies. They were comfortably furnished and for an extra fifteen shillings a week one could have a piano on board. It marked the start of pleasure boating as a major activity and a variety of wherry yachts and pleasure wherries were available for hire.

Today, when cargo carrying has come to an end, pleasure craft dominate the Broads, but these days they are mostly motor cruisers. The rules, however, remain the same as they were a century ago; power gives way to sail. Not all hirers seem aware

***Albion* at** Bungay Staithe in her working days when she was one of a fleet of wherries owned by the Bungay maltsters W.D. and A.E. Walker.

An old postcard showing a pleasure wherry on the Norfolk Broads. The derivation from the working wherry is obvious, even down to the traditional white triangle painted on the bows. The sail, however, has changed from black to red.

of this. We were sailing one day with a brisk wind behind us and making good speed. A motor cruiser was sat at the side of the bank and for some reason, the hirer decided that as we got close would be the ideal time to make a U-turn in the river in front of us. Our skipper at once saw the danger and ordered the sail dropped immediately, but momentum carried us relentlessly on – wherries are not fitted with brakes or reverse gears. There was no room in the narrow river to manoeuvre and we hit him broadside on and we continued on our way with the launch stuck on the bows, until everything came to a halt at last. It turned out that it was a family hiring and the teenage son had been at the wheel – to say his parents were unimpressed would be an understatement. One thing was certain – their holiday was over thanks to a large hole in the cruiser's hull. If nothing else, it was a testament to the wherry's stoutness, the only damage being some paint chipped off in the collision. Happily, such events are rare, and *Albion* continues to give delight to all those fortunate enough to sail aboard her.

CHAPTER THREE

The Thames Barge and More

The Thames has been an important trading route since at least Roman times and until recently, London was the country's most important port. From medieval times, however, the river had been divided into two distinct sections, above and below London Bridge. The first bridge, built by the Romans, did not survive their departure from Britain, and the first medieval bridge had a central section that could be lifted to allow boats to pass through. But what we now think of as the old London Bridge, begun in 1176, had no lifting section and the multitude of arches differed in span from the narrowest at 10ft to the largest at 33ft. The water raced through these narrow openings, so that on a fast-flowing ebb tide there could be a difference of 5ft or more between the water on one side of the bridge and the other. Inevitably, taking a vessel under the bridge was a hazardous task, and accidents and even fatalities were far from uncommon. But, for larger sea-going cargo ships, it was simply an impassable barrier; they could only gather downstream in the Pool of London. And that was the situation for the next 600 years, until it was replaced in 1831.

For centuries, there were no enclosed docks along the river. In the early medieval period, visiting vessels would haul up alongside a quay, where they would settle on the bottom at low tide. But by the fourteenth century, as vessels became bigger, they mainly anchored instead in the centre of the river, almost filling the Pool. A writer noted in 1796 that a stranger visiting London for the first time, would 'lament the contraction and embarrassment of the stream from the tiers of ships moored in the midst of it, and reaching from each side, so as at some times scarcely to afford a passage, and liable to various injuries'. These ships had to be served by a fleet of lighters and barges for loading and unloading cargo. A definite split developed between the watermen, who carried out this work, the lightermen, and those who worked above London Bridge, carrying cargo upstream, the western bargemen.

The barges and lighters of the Pool moved with the tide, controlled by sweeps, when loading and unloading the ships. But the barges below the bridge also made longer voyages, down the Thames and round the coast to travel up the Medway and along the Lea. They were not elegant craft, with basic hulls, straight sided and with flat, sloping bows and sterns, rather like an oversized punt. Like the keels, they had a single mast and a square sail. The western barges travelling up the Thames faced special problems. The sails were only usable with a following wind and for the rest of the time they were bow hauled, often by gangs of men as can be seen in the illustration (below) showing craft sailing downstream at Windsor, while another barge is being hauled by a gang of men on the towpath. There were other difficulties facing the western bargemen, particularly the wretched condition of much of the river as far as navigation was concerned. John Taylor, the waterman poet, wrote a verse description of a journey down the Thames, which was published in 1632, in which he listed the hazards met along the way. He set off from Oxford, and was soon in trouble.

A busy scene on the Thames at Windsor in the seventeenth century. The barges coming downstream are typical of the early Thames barges, box-like with sloping bows and a simple square rig. Those moving upstream are being hauled past a mill leat by a team of horses.

> At Abingdon the shoales are worse and worse
> That Swift ditch seems to be the better course.

He reached the flash lock at Sutton. Vessels coming upstream had to be winched through. As Taylor noted, this could be a difficult operation.

> Then Sutton locks are great impediments,
> The waters fall with such great violence,
> Thence downe to Cullom, streams runs quick and quicker,
> Yet we rub'd twice a ground for want of liquor.

Things got no better as he went on.

> At Clifton there are rocks, and sands, and flats,
> Which made us wade, and wet like drowned rats.

It may not be the world's greatest poetry, but it does paint a vivid picture of barge travel on the river in the seventeenth century. Sometimes, the water levels were so low in this period that a large loaded barge would be unable to get far upstream, and had to transfer the load to two or three other barges that with less weight on board could float over the shallows. It was not really until the eighteenth century, when apart from a few flash locks on the Upper Thames, the river had pound locks – the familiar locks we all know today with two sets of gates – and water levels were preserved and dredging regularly carried out. Even when the pound locks were constructed, there were still problems on the river. A complaint to the Thames Navigation Commission in 1820 reported an unhappy series of accidents:

> The barge *Emma*, laden with copper, and proceeding down the stream of the Thames, with all the expedition that could be used, was sunk on the 21st of February last on a shoal near Bell Weir Lock … the same barge on her return being laden with grocery, &c. was again sunk on the 20th of March last, a little above Boulters Lock, having first been dragged over a shoal between Windsor and Maidenhead!!!

Navigational hazards were not the only problems facing Thames barge owners. When William Darvell of Maidenhead tried to take his Thames barges up the Kennet in 1725, he received this threatening letter from the local bargemen. The original spelling has been left as written:

> Mr Darvell wee Bargemen of Redding thought to Aquaint you before 'tis too late, Dam You, if y. work a bote any more at Newbury wee will Kill You if ever you come any more this way, wee was very near shooting you last time, wee went with to pistols, and was not too Minnets too Late. The first time your Boat Lays at Redding Loaded, Dam You. Wee will bore holes on her and sink her so Don't come to starve our fammeleys and our Masters …so take warning before 'tis too late for Dam You for ever if you come wee will doo it – from wee Bargemen.

I could find nothing in the records to say if Darvell heeded the warnings or not.

We are still a long way from what we now think of as a typical Thames barge. Its distinguishing feature is the spritsail. This is a fore and aft sail, and the sprit that gives it its name is a spar that runs diagonally, from a point near the foot of the mast to the peak of the sail, the far top corner. The lower far corner nearest the stern, the clew, is controlled by the sheet. Two ropes attached to the top of the sprit, the vangs (pronounced 'wangs'), lead down to the deck and are used to control the movement of the sprit. The bottom of the sprit rests in a rope collar on the mast, the snotter. The latter is prevented from slipping down the mast by a standing lift or stanliff, a rope running down from the masthead. When not needed, the sail is braided, pulled in to lie against the mast and the sprit.

It is generally agreed that this type of rig originated in the Netherlands and the earliest illustration we have shows quite a small craft, with a single sail and tiller steering. It is a miniature, illustrating the voyage of the Saints in the Low Countries and appears in the *Heures de Turin* of c.1420. Other illustrations of a later date show more complex and bigger vessels with spritsails, but none feature in English illustrations over the next two centuries.

It seems that the idea was introduced into this country during the reign of Charles II (1660-85) when the Dutch presented him with a 'jaghteship' or 'fast ship' – a name that became Anglicised to 'yacht'. Various vessels of this type were built with spritsails over the following years. So, vessels with fore and aft sails were becoming more and more common by the end of the seventeenth century, but on the Thames development was hampered by the ungainly hulls of the barges.

One of the earliest representations of a spritsail barge comes from a work of 1768 by the Swedish master shipbuilder Frederick Henrik. Although published in Sweden, simultaneous translations appeared in several languages, the English version appearing as Chapman, *Architectura Navalis Mercatori*, and among the vessels that were described was a Kent chalk boat. This was presumably used to bring chalk from Kent down the Medway and up the Thames to London. She was a comparatively large vessel by the standard of the time, 56ft long overall and 15ft beam, with fine lines – sloping side instead of the slab like

A Thames barge of the nineteenth century beside the old Hungerford Bridge in London. This is a sprit sail barge, with the main sail furled to the sprit and the mast. There is also a jib sail.

sides of the other barges of the area. She had considerable 'swim' at the bows – that is, the flat bows rose at an acute angle from the keel. She was flat bottomed and, as a vessel designed for coastal voyages, was equipped with lee boards. Her single mast carried a spritsail and a triangular foresail. The mainsail, as on *Albion*, was controlled through a sheet block running across a horse, though constructed of wood not iron. Steering was by tiller. In the illustration there is no indication of hatches covering the hold. Possibly they were simply omitted from the drawing, but if this really was an open boat, then any sea trip must have been decidedly tricky. While not quite like the Thames barges we see today, we can see that this represented a major advance over the simple craft mostly used on the river.

By the early years of the nineteenth century, barges were getting bigger and longer, which tended to make them more difficult to handle when tacking. To help overcome this problem a short mast was added at the stern – the mizzen – usually set with a small gaff-rigged sail. The sail was originally stepped to the top of the rudder. When the tiller was put hard over the wind would catch this sail and help push the stern round. This arrangement lasted until the tiller was replaced by the wheel for steering, when the mizzen was moved on to the stern deck. The other major change that came over the years was the replacement of the punt-like swim bows with more rounded bows. The larger barges also added a topmast and topsail above the main, with a jib sail from the mast to the bows or with more than one, attached to a bowsprit.

One of the great spurs to development came in 1863 when Henry Dodd, a contractor, initiated an annual race for barges. It was a sporting event, but Dodd believed that it would lead to overall improvements in design as builders vied for the honour of winning and the prestige that it brought. At the start there were two classes of barges. For the third race of 1865, the first class were the topsail barges, with a maximum of 100 ton and there were a series of prizes: an £18 cup to the owner, and £10 10s to be shared among the crew; £12 cup and £5 5s to the second; and £8 cup and £3 3s to the third. Slightly smaller prizes went to the second class, the stump-rigged barges, without topsails, up to 80 tons. The barge race tradition continues to this day, but now with a whole series of races round the south east coast, and the categories have slightly changed; the difference now is between barges with or without sails set on a bowsprit. The big difference, of course, is that the barges no longer trade and carry cargo, and crews are amateur. They do not get payment for winning in cash, but there are still prizes handed out. I am the proud possessor of a winner's mug from a match on the Medway. I have to admit, however, that the barge I crewed was not a winner – but there were more mugs made than there were crew in the winning barge. I bought it – and have yet to crew on a winning barge. That, however, does nothing to spoil the pleasure of taking a part in one of these splendid events.

Most barges these days have motors, but before the start of racing, the order comes to shut off the engines, and the barges start manoeuvring to be in just the right place when the gun is fired for the off. The worst mistake at this stage is to find oneself either over the line before the start or facing the wrong way. It all depends on the skill of the captain, and many barges have old barge skippers at the helm for the day. I was lucky enough to sail with one of these gentlemen, who told me how, as trade became worse and fewer cargos were available, he was reduced to working the barge single-handed. It scarcely seemed possible. On match days, there are generally more hands than there would have been on a barge in its working days, and everyone has a specific job to do. I remember a match when I was allocated the task of dealing with the lee boards. As the vessel tacked, the board on one side had to be winched up out of the water, a task that got harder as it emerged and lost buoyancy. Dropping the board on the other side was, of course, a lot easier. But having to do this, manage the sails and cope with steering single handed seemed to me to be all but impossible. It is not just the pleasure of being aboard a sailing barge that makes the occasion, but the pleasure of seeing so many of these magnificent vessels all out on the water at the same time. They are days to remember.

The Thames barge as it developed changed over the years. The lines tended to become somewhat finer and the vessels larger. Wooden hulls gave way to steel and the tiller steering was replaced by the wheel. Working the craft was, in some ways, similar to the wherry, with the mainsail sheet running

In order to encourage the development of the Thames barge, a series of annual races were instituted in 1863. This illustration of the fifth match of 1867 by Josiah Taylor shows the barge *Renown* in the lead.

down to the horse, but with a separate horse for the foresails. When coming about, the mainsail swings first followed by the foresail. Some barges were specialised. One example is the hay barge, which provided an essential service when all the traffic on London's streets relied on horses. The barges themselves were slightly shallower than the other barges to allow them to reach farms further inland. There they loaded with hay, stacked up to ten feet high. This meant that the helmsman had no way of seeing where he was going and had to rely on the mate, perched on top of the stack, for directions.

Thames barges had one of their finest hours during the Second World War when sixteen of them joined the flotilla of small boats heading for Dunkirk to rescue the troops on the beaches. They were towed across by powered craft, but with their comparatively shallow draught were able to get close to the shore to pick up the men. This put them in the heart of the most dangerous zone, but the captains and their crews never flinched. Several vessels were damaged and abandoned, though one of them, the *Ena*, somehow broke free and with no one aboard sailed back to English waters. One of the barges that did make it back safely with the crew was

One of the most important cargoes for London was hay for the vast array of horses who provided power for London's traffic before the arrival of motor vehicles. The illustration shows a hay barge opposite the old shot tower that once stood near the present Royal Festival Hall. It is obvious from the illustration that the steerer could not see where he was going and had to rely on the mate for instructions.

Pudge. She was taken over the Channel by tug, and on nearing the beaches, an explosion lifted her clear out of the water, but as her skipper laconically noted, 'came down the right way up'. Built with a wooden hull in 1922, she ceased trading in 1968 and is now owned by the Thames Sailing Barge Trust. At the time of writing, the Trust are putting the finishing touches to a restoration programme, which should see her out on the water to celebrate her hundredth birthday. The Trust's other barge *Centaur* is even older, constructed in 1895, and two years later took first place in the Harwich barge match. She was a big coastal barge, able to make extensive journeys round the British coast and over to Europe – she was a regular visitor to Calais. She too was made ready for the Dunkirk rescue but was damaged by a tug at Dover and had to stay behind for repairs. The Trust provide training opportunities to ensure that the craft of sailing these fine vessels is kept alive and do regular charter work and when *Pudge* is fully restored both vessels will be available for chartered cruises. They are only two of a number of barges still afloat, many of which also offer trips to the general public, including on match days.

The Thames Sailing Barge Trust's barge *Pudge*. This clearly shows the sail arrangement, with two jibs and the mainsail extended by the sprit running at a diagonal from the foot of the mast and a second small mizzen spritsail at the stern. She was among the barges that took part in the evacuation of troops from Dunkirk.

So far, we have been looking at a variety of different sailing barges, each with its own individual rig, in some detail. But every river system had its own variations and the following descriptions should give an idea of the rich variety of craft that could once be found on Britain's inland waterways. I have chosen craft again where there are surviving examples to be seen. The first example is *Shamrock*, an elegant vessel built in 1899 at Devonport for an inelegant trade. She was used to carry animal manure from the west bank of the Tamar to Plymouth. She is only 57ft 6in long, but wide in the beam and has a shallow draught. When grounded she could stay upright, which meant that the carts of manure could be brought right alongside, straight from the farms. She had no keel, but she had a centre board that could be dropped into the water amidships – a device that was common on sailing dinghies. She was sold in 1919 and converted into a vessel capable of making short sea

passages, thanks to the addition of a false keel. A bowsprit was added to take two foresails and hatch coamings were raised. She has been fully restored to her 1919 condition, with two masts and fore and aft gaff rig. Unlike most restored craft, she was given not just sails and rigging that looked like the original, but original materials were used. Sails are canvas and hemp rope was manufactured for the rigging. As I discovered, handling hemp is very different from working with the artificial fibres in general use today. The lines seemed very coarse and hairy and the palms of my hands itched for days – but that is probably a very landlubberish reaction.

One of Britain's busiest rivers was the Severn, so important that it was known as the King's Highway of Severn, and the sailing barges that traded on it were known as trows. They not only worked the Severn as high as Ironbridge but also

The Sailing Barge Trust's other barge, *Centaur* was built in 1895. She is a wooden coastal barge, bigger than the standard river barges and able to trade with ports around Britain and across the Channel.

The ketch rigged Tamar barge *Shamrock* at Cotehele Quay, restored by the National Trust with the National Maritime Museum.

made trips up the Wye and out into the Bristol Channel. Like the Thames barges, they saw considerable changes over the years. The illustration (below) shows the trow at the end of the eighteenth century, with a single square-rigged mainsail, and the ability to attach foresails to the bowsprit. Like all river craft she could drop her mast, as can be seen with the other trow in the picture. One distinctive feature was the D-shaped transom stern that was to remain a feature of the vessels throughout their long period of development. Some trows of this period also carried a topsail. There was a busy traffic all along the river and into the canal system. One example was the trade in coal from the mines of the Forest of Dean to the woollen mills of Stroud as they converted to steam power, travelling from the river and along the Stroudwater Navigation.

By the middle of the nineteenth century, more and more boats needed to travel the estuary waters, which required some major changes. From being single masted and square rigged, they developed far more complex rigging, with a gaff rigged mainsail, topsail and foresails together with a short mizzen. Traditionally, the trows had needed to be flat bottomed to enable them to travel far upstream, but this presented a problem when

Severn Trows at Gloucester at the end of the eighteenth century, engraved from a drawing by G. Catton. These are comparatively small craft compared with later trows. The vessel to the right with its mast dropped is being bow hauled under the bridge.

A trow on the Stroudwater Navigation. Unlike the trows in the previous illustration this one is fore and aft rigged. The former Ebley woollen mill in the picture still stands but is now offices and the house no longer exists.

sailing in the open waters of the estuary, so a keel plank was added. This was generally a substantial timber 28ft long and 2ft wide. At each end of the plank were chains attached to brackets. When the plank was lowered into the water by one set of chains that were made fast, the other chains were passed round the stern, and used to haul the plank into the correct position, when they too were made fast. These later trows usually had built up coamings and covered hatches. By the end of the nineteenth century, trows were being built with iron hulls.

The last surviving trow is *Spry*, rigged with a single mainsail and two foresails, but no topsail or mizzen. She is 71ft long and 18ft beam at the widest and carried up to 40 tons. She no longer sails but is housed in a special shed at the Ironbridge Gorge Museum.

The Stour barges are familiar from Constable paintings, stumpy, wide beam day boats, just 47ft long and 19ft 9in wide. The vessels always travelled in pairs, and both could fit into

Trows were not limited to the Severn and adjoining canals. This pair are trading on a somewhat romanticised River Wye above Chepstow.

the 95ft long locks. They did, however, have a unique way of working as was described in a book of 1916.

The bow end of the fore boat has a deck large enough to provide standing room for a horse. Whenever the towpath crosses the river, a small jetty is built out into the stream

DANKS, VENN & CO.
Having taken up the Business lately carried on by
BELSHAM & Co.
RESPECTFULLY INFORM THEIR FRIENDS THAT THEIR
LOCK-UP TROWS,
REGULARLY SAIL TWICE A WEEK,
TO AND FROM
Bristol, Glo'ster, Worcester, Stourport & Bewdley;
And by their own Boats to Stourbridge, &c.
By which every description of GOODS, particularly Wines and Spirits, are safely and expeditiously conveyed for the undermentioned Places;

LUDLOW	WOLVERHAMPTON	BIRMINGHAM	MANCHESTER	AND ALL PARTS
SHREWSBURY	THE STAFFORDSHIRE	COVENTRY	AND ALL PARTS	OF THE WEST
KIDDERMINSTER	IRON WORKS AND	LEICESTER	OF THE NORTH	SWANSEA
STOURBRIDGE	POTTERIES	DERBY	BATH	AND
DUDLEY	WALSALL	LIVERPOOL	EXETER	SOUTH WALES

FOR PARTICULARS OF FREIGHT, &C., APPLY TO

WM. BIRD, Wharfinger,
DANKS, VENN & CO.   } Stourport;
JOHN DANKS, Wharf, *Birmingham*;
CROWLEY & CO. *Wolverhampton*;
HOOD & WALL, Diglis, *Worcester*;

ANTWIS & STURLAND, Castle Fields, *Manchester*, or Duke's Dock, *Liverpool*;
W. KENDALL, Wharfinger, *Gloucester*;
T. Y. VENN, Wharfinger, 6, Tontine Warehouse, Quay Head, *Bristol*.

D. V. & Co. request the favor of their Friends to be particular in directing Goods UPON THE PACKAGES to be forwarded by their Trows.

[MARY NICHOLSON, PRINTER, BRIDGE-STREET, STOURPORT.]

This advert shows the versatility of the trows, which traded high up the Severn and down the Bristol Channel to South Wales. The vessel in the crude illustration does show how the trow could develop into one of the largest of all Britain's sailing barges with mainsail and topsail, jibs and a gaff rigged mizzen.

A sketch by John Constable of men loading a lighter on the River Stour. These were very basic open, horse-drawn boats with an open hold and small steering position in the stern.

corresponding to another on the further bank – not opposite but a few yards up or down stream. When a crossing is to be made, the horse ceases to pull and goes to the pier head. The boat is steered close alongside and the horse leaps on board, some litter having been considerately placed on the deck to afford him foothold. The boat is then steered to the opposite pier, where the horse leaps out ready to begin work again.

There were other regions with their own distinctive barges, including the 'flats' that traded on the Mersey and Weaver. It is not necessary to go into every detail except to say that they adapted over the years to changing circumstances. Their size, however, was to turn out to have special significance. In general, the flats had a maximum length of around 70ft and a breadth of 15ft. We shall see exactly why this was significant in the next chapter.

A Mersey flat on the Sankey Brook Navigation, passing under the viaduct of the newly opened Liverpool & Manchester Railway. The arrival of the railway proved to be the death knell for the Navigation and visiting this spot today there is little to show there was ever a navigable waterway here at all.

Chapter Four

The Canal Age

There is some disagreement about when Britain's Canal Age really began. Technically, the first canal to be built was the Sankey Brook Navigation, which was approved by an Act of Parliament of 1755. It took its name from the fact that the stream provided water for the twelve-mile-long canal, from St. Helens to Widnes on the Mersey. It had eight single locks and two double locks, each of which was 15ft wide and were specifically built to take the Mersey flats described in the last chapter. The reason it has never received much attention as a pioneering work lies most probably in the name. 'Navigation' was generally applied when an existing navigable river was improved by the addition of locks. But the little Sankey Brook was never navigable, and never could have been made navigable, hence the confusion. Incidentally, the name 'Navigation' gave us another common word. At the start of the canal age, the men who did the actual work were known as 'navigators', later shortened to the more familiar 'navvy'. At the time, the Sankey Brook raised no general interest, simply because it was seen, erroneously, as no more than an extension of an existing system. The same could not be said of the canal that also lays claim to starting Britain on the way to creating a canal network throughout the country.

This part of the story really begins in France. Francis Egerton, 3rd Duke of Bridgewater, was a weakly child who was not expected to live to inherit the dukedom. But when he did survive, he was hastily shipped off to Europe, with a tutor, to take the fashionable Grand Tour, an essential rite of passage for a young nobleman in the eighteenth century. The tour started badly in Paris when the young man, not surprisingly, showed more interest in dubious young women than in works of art, so was taken on to Italy where he did what he was supposed to do, purchased several works of antiquity, which were packed and sent back to England. Popular history has it that the crates were never unpacked. But on his travels, he did see something that interested him. He was greatly impressed by the seventeenth-century French canal, the Canal du Midi, an imposing piece of engineering, with locks, a short tunnel and something very new, an aqueduct. Back in London, he fell for a society beauty, the Duchess of Hamilton. However, her sister was involved in a minor scandal, and the young man informed her that she must never see her sister again. She, very reasonably, refused this outrageous suggestion and the affair was over. The Duke returned to his estates at Worsley near Manchester and swore never to have anything more to do with the opposite sex. Instead, he decided to improve his estates, which already contained coal mines at Worsley that he wished to develop.

He realised that one way to make more money from the mines would be to cut the transport costs, by means of water transport. His idea was to build a short canal to join the Mersey and Irwell Navigation, but that important company wanted nothing to do with it and refused him the right to join the system. It was then, in consultation with his agent, John Gilbert, that he came forward with a far bolder plan. He would build a canal all the way from the mines to the centre of Manchester, and if he could not join the Irwell, then he would leap over it in an aqueduct. The idea was scorned by many, including some practical engineers who should have known better, and one remarked that although he had heard of castles in the air, this was the first time he had come across anyone actually planning to build one. But, of course, the Duke had seen a navigable canal crossing a river in France and knew it was possible. All he needed now was an engineer to put his ideas into practice. The man he chose – James Brindley – had begun his working life as a millwright but had also previously worked on a drainage scheme for a colliery. It was an unlikely trio – a land agent, a 22-year-old Duke and a millwright, none of whom had ever had any practical experience of canal construction. But they succeeded and Britain was presented with its first navigable aqueduct at Barton, where the Bridgewater Canal soared above the River Irwell. Barton aqueduct became a tourist attraction and made the little canal famous, but its true importance was measured in more attractive terms for potential investors in canals – the price of coal from

The Bridgewater Canal led deep into the heart of the coal mines at Worsley and was used by very narrow boats known as starvationers, because they had prominent ribs in the frame. One of these vessels is being used by inspectors checking the underground system.

the Worsley mines when sold in Manchester was halved.

What most people did not know was that there was an even more extensive system of underground canals, reaching deep into the mine. Special boats, just 4ft wide, were built, which could be loaded directly at the coal face, and then brought out into the open at Worsley Delph, the start of the wide canal. There three or four could be linked together and pulled along the canal by a horse. These thin vessels had prominent ribs, which gave them the popular name 'starvationers'.

At Manchester, the coals arrived at the foot of Castle Hill. For people buying coal for their houses, it was a long trudge up with their load, so to overcome that the canal was extended into the hill. A shaft was then sunk down to water level and a hoist erected at the top to pull up the coal.

The canal was completed in 1761, and the following year was extended to the Mersey at Runcorn. Brindley was again the engineer in charge. He developed his own method of canal construction. Wherever possible, he avoided obstacles. He preferred to go round them, rather than over them in locks or through them with cuttings or tunnels, even if it meant that his canals tended to wander across the countryside like a natural river. But there was no escaping the fact that when the canal reached the Mersey there was a considerable difference in height between the canal and the river. In all, ten locks were needed to bridge the gap, and Brindley took the obvious decision that they should, like the Sankey, be able to take the flats from the river. So they were built to take vessels 72ft long and 15ft wide. In the well-known portrait of the young Duke, he points proudly at the barges being drawn across Barton aqueduct, while a team of men labour to haul one on the Irwell down below (page xxx).

THE CANAL AGE • 49

The young Duke of Bridgewater pointing proudly at the Barton aqueduct carrying his canal over the Irwell Navigation. The illustrator has emphasised the benefits of the canal by showing a string of barges serenely travelling along the canal, while down below a large team of men strain to heave a boat on the river.

It might have been thought that this would have set a standard for future canal construction, which would have meant that, as the system grew, the same boats would be able to travel widely throughout it, in much the same way that when George Stephenson first built a railway with tracks 4ft 8½in apart, he began what would eventually become accepted as the standard gauge for Britain. Thanks to the success of the Bridgewater Canal, there was a flurry of new canal proposals put forward and Brindley was very much the man of the moment. There was every reason to think that he would continue with the same size of locks and thus make all his canals available to similar sized barges. But among the first of the new canals, which obtained its Act in 1766, was the Trent & Mersey. And here, Brindley was faced with a problem, and one he couldn't go round – Hardcastle Hill. He could not go over the top, for there was no water on the summit to feed the locks. The only option was a tunnel, but it would need to be a mile and a half long. He felt that to drive a tunnel that would take 15ft wide boats was simply too much for the available technology, so he decided to halve the width. This is a huge saving. It does not reduce the amount of material by half. If you think of the tunnel as being basically a round tube, which it isn't exactly, but it is near enough to make the point, then the area of the tunnel opening is given by the square of the radius multiplied by the constant pi. That means the area of the narrower tunnel mouth is not a half but an eighth of the wider one, which is indeed a big difference. Given that it actually took eleven years to complete even the narrow tunnel, it was probably a rational choice.

Having taken this momentous decision, he could have said to himself that, although only boats of around 7ft width can pass through my tunnel, I could still build locks to the wide dimension so that they could be used by two boats at a time. But he didn't; he designed locks to the same narrow size. And now he decided that this would be the pattern he would follow for other canals for which he was engineer. There were four of these that together formed a cross that linked four great rivers, Trent, Mersey, Severn and Thames: the Trent & Mersey; the Staffs & Worcester; the Coventry; and the Oxford. And at the heart of the system was the Birmingham Canal that served that rapidly developing industrial centre. That is how it came about that the narrow boat, roughly 70ft long and 7ft beam, became the most common vessel to be found throughout most of the English canal system, and in particular in the Midlands.

Not all narrow boats, even in the early days, were the same. The illustration (below) shows two very different boats. On the left is the most common form of narrow boat, with a cabin in the stern – and obviously the stove is on, as smoke can be seen coming out of the chimney. On the right is a 'joey' or day boat. These were common on the complex of canals centred on Birmingham. As the name suggests, they were not used for long journeys, but only for local traffic, so there was no need for a cabin as the crew went home at the end of the day. They were comparatively crude, with rounded bow and stern, though those terms hardly apply as they were designed symmetrically, so could be towed from both ends. The horse drawn joey was still in use in the region until well into the second half of the twentieth century, when they were used for moving scrap metal around the system. But there is good evidence, that craft similar to these were normal throughout the early years of life on the canals.

A pair of narrow boats on the Birmingham Canal passing under Galton Bridge. Both are horse drawn, but the boat on the left has a cabin for the crew who can make long journeys, while the boat on the right is a day boat only working in the Birmingham area so has no need for sleeping accommodation.

Mrs Rose Skinner standing by the back cabin of the horse boat *Friendship* that she and her husband Joe worked for many years on the Oxford Canal. The boat is now preserved as part of the National Waterways Museum collection.

Canal companies were at first forbidden by law from running their own boats, so that was left to carrying companies who had one or more boats and hired crew to run them or to 'number ones', who worked and owned their own boat and had the horse to pull it. The last of the number ones were Joe and Rose Skinner, though Joe preferred mules to horses. They worked almost exclusively on the Oxford Canal and when Joe died, Rose was given accommodation on the land, but each day she would go down to Hawkesbury Junction where their old boat, *Friendship*, was permanently moored and sit in the back cabin, where she always welcomed visitors for a chat and would bring out her well-worn photos of their working life. Rose, too, has died now, but the old boat has a permanent home in the boat museum at Ellesmere Port.

The Skinners, however, had the advantage of a cabin on their boat; the early canal boatmen rarely enjoyed such a luxury. There were inns built alongside the canals which would have provided accommodation, but not everyone could afford them. A report in the *Birmingham Daily Mail* in 1875, described the life of crews who worked on open boats when they had to make long journeys away from home:

> They have no shelter, and often sleep one night on board, doing the best they can. Their fire is a huge open circular grate, such as we see at night on roads under repair, and it seems to me that the approved mode of taking a siesta is to lie flat on the back, with boots as near the fire as may be convenient.

There is some uncertainty about just when and how boating families left their conventional homes to live full time on their boats. One plausible theory, put forward by Harry Hanson in his book *The Canal Boatmen*, is that the arrival of the railways brought stiff competition and the move was essential to reduce costs. Where before two men would have worked a boat, each perhaps with a family to support and a home to maintain, now all the family could help out and could make their home afloat. There is some evidence to support the idea that the railways hit canal transport quite severely. One important carrying company was Pickford's.

One of the earliest mentions of the company came in an advert in the *Manchester Mercury* for James Pickford 'The London and Manchester Wagoner' to inform customers that they had moved the London base from Blossom's Inn to the Bell Inn. At this time, he was offering a regular service twice a week, which would have involved the use of heavy wagons pulled by a team of horses. By 1772, the business was being run by James' son Matthew and he bought a single boat, seemingly as an experiment to see if it was as profitable as their road haulage business. As time went on and the canal system spread, they established an important trade. At their London wharf at City Road Basin, they established a regular system, where in the morning boats brought in cargoes for unloading and distribution around the city, while in the evening, carts brought loads from various collecting points, ready for despatch. By the early years of the nineteenth century, they had 80 boats and 400 canal horses. Theirs was a super-efficient system. They had fly boats, which ran to a fixed schedule travelling day and night, using relays of horses and change crews. Theoretically, they had precedence over other craft, though claiming their rights over another boat must have led to some interesting arguments. Then the railways came along. Pickford's began sending goods by rail, steadily reducing their fleet. In 1847, they announced that they would cease the canal trading the following year. If a highly efficient organisation such as this found it difficult to compete, one can see why reducing costs would be vital for all carriers, and that the family boat would have been one answer to the problem.

A pair of Pickford's narrow boats on the Regent's Canal, looking towards the entrance to the Islington tunnel. The man in the tall hat is a canal official and is holding a gauging rod, used to measure how deep a boat is sitting in the water, from which the weight of cargo can be calculated and appropriate tolls collected.

Although we do not have exact information on when the change came about, we can say that, by the middle of the nineteenth century, the narrow boat as we know it today was firmly established. There were small variations in the craft, depending on the individual builder, but the general pattern remained the same, the dimensions being standardised at approximately 70ft long by 7ft beam. Starting at the bows, is the stem post, to which the side planks are attached. The bow curves sharply into this post for roughly 7ft, then the sides remain straight until they curve again towards a similar upright post in the stern. The first few feet of the vessel are decked over to create a storage space, beyond which is the open hold. This is one of the features that distinguishes narrow boats from barges. The holds of the latter are generally covered by hatches. At the start of the hold is a triangular wooden board called the 'cratch'. There are three posts, evenly spaced, between the cratch and the front of the cabin, which carry top planks over the hold. The first of these is extended upwards through the planks as the towing post, to which the tow rope is attached. Some cargoes, such as grain, have to be protected from the elements, and on these occasions, the boat is 'sheeted up', canvas covers spread over the top planks and secured. At the after end of the hold is the cabin, and beyond that an open space for the steerer, and the wooden rudder and tiller. The tiller fits into a slot in the rudder post and is curved down when the boat is on the move but taken out and replaced to point upwards when moored, to allow more space in the cockpit area and give easy access to the cabin.

It is obvious that in such a confined space, perhaps no more than by 7ft by 8ft, organisation and adaptability are the key to life in the cabin. Down one side would be the side bed, a double bed, though a distinctly cramped one not much more than 3ft wide. Opposite that was a cupboard, the door of which could be hinged down to make a table. There were cupboards and drawers to stow everything away and a door forward that gave access to the hold. The most prominent item was the stove or in later years a range, which was used for both heat

This old wooden butty has reached the end of its working life and is being cut up. The photograph does, however, provide an opportunity to see the shape of the hull as it tapers from the straight side towards the sturdy stern post.

A couple on a Fellows, Morton & Clayton butty. He has the typical company uniform and she is wearing traditional boatwoman's dress and headgear. The photo also shows the traditional decoration of roses and castles on the cabin doors and sides.

and cooking. The cabin itself was panelled, but little of that was to be seen, for every available surface was covered with decorative ornaments. Lace plates were a particular favourite, and brass knobs were attached to anything that needed to be moved. It was a matter of pride to keep everything highly polished.

One very distinctive characteristic of the narrow boats was the bright decoration on every available surface. There were geometric patterns on the hull and stern post, while panels such as cabin doors and external walls had romantic scenes, usually roses and castles. Some have suggested that the style is reminiscent of the Roma caravans, but there is not much evidence to suggest that boating families had gypsy ancestry. My own theory, which I have no way of proving, is that this was a reaction to what the families had lost when they moved permanently on board – a cottage and its garden. Many boatyards had specialist painters, who not only provided the exotic scenes but were responsible for the ornate lettering, identifying the

The interior of the back cabin of a narrow boat with the black-leaded stove and a prolific display of lace plates.

boat. The decoration also appeared on many items of daily use such as the Buckby can that was used to collect fresh water.

There were several variations on the basic narrow boat, depending on the yard that built it, but one type was very distinctive. Although in general narrow boats had open holds, there was one exception. Thomas Clayton of Oldbury had a very

A typical horse drawn narrow boat or butty, with the cargo hold partly sheeted over.

specialist trade that only began in the nineteenth century with the development of gas works. One of the by-products was coal tar, which was widely used as a preservative, especially for railway sleepers. It was later used for road surfacing as an ingredient in tarmacadam. The Clayton boats were, in effect, tankers, with a flat deck over the hold. The Clayton firm would eventually amalgamate with a company founded by James Fellows of West Bromwich in 1837, which was joined in 1876 by Frederick Morton and in 1889 became Fellows, Morton & Clayton, one of the largest and most important carrying companies on the canal.

The boatmen on the canals did not face the same challenges as those of the sailing barges, but they had their own problems to overcome. One of these came when they reached tunnels, many of which had no towpath. The answer was to leg them through, by two men, lying head to head, facing in opposite directions, placing their feet against the tunnel sides and walking the boat along. On wide tunnels, such as those of the Grand Union Canal, special boards had to be run out from the sides of the boat to

enable them to reach the tunnel sides. Meanwhile, the horse would have to be led over the top to rejoin the boat at the far side. This was work that was often left to one of the children. Nell Cartwright was just thirteen years when she was given the job of taking the horse over Braunston tunnel at one in the morning to meet a deadline the next day. She was terrified at first but the horse nuzzled against her. 'I thought to myself, "he's telling me not to be afraid", and when I got to the end of the tunnel, I was as brave as brave.' In her later teens she joined her brother in legging the boat, for which she was given pocket money of a farthing a fortnight; if she saved up for two months, she would have had a whole penny to spend. She was eighteen before she finally got a wage for the work she did. On some of the longer

Not all narrow boats were the same. Thomas Clayton of Oldbury specialised in carrying the waste products of the gas industry, and had special tankers built, recognisable by their flat deck. The horse has stopped, the tow line is slack and the boat is slowly edging into the lock.

tunnels, such as Standedge on the Huddersfield Narrow Canal (5,415 yards/4,951 metres), professional leggers were used.

The horse was as important as the crew in ensuring that a trip ran smoothly. Most families kept the same horse for as long as it was able to continue. After some time at work, the animal would recognise, for example, that a lock was a place where it should start slowing down. The narrow boat, unlike a modern pleasure boat, had no reverse mechanism, so bringing it to a halt without causing damage either to the lock or itself required a certain amount of skill to get a line round a bollard to stop the vessel before it crashed into the far gates. Unless the horse had already stopped, the task would be that much harder. The more difficult task for the horse was getting the boat under way. This could be

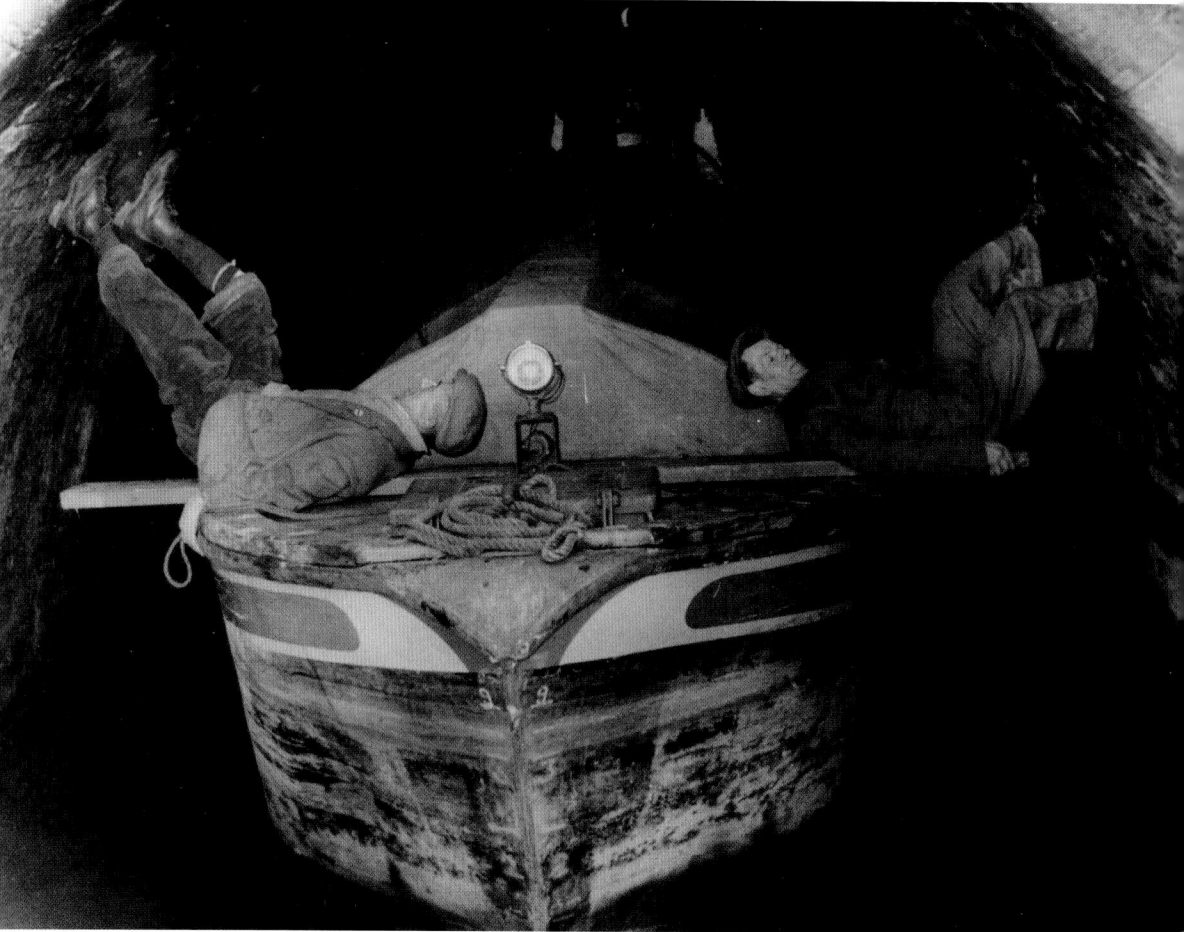

Legging a boat through Barnton tunnel on the Trent & Mersey. On some of the longer tunnels professional leggers were employed.

helped by the use of a pulley on the mast head and by looping the tow rope round a bollard ahead of the gates. The horses also learned to look after themselves, nodding their heads to one side as they passed under the curved arch of a bridge. Some parts of the journey must have taken some getting used to. One can imagine the first sense of fear when crossing an aqueduct, especially one as spectacularly long and high as Pontcysllte near Llangollen. Most horses arrived at canal work after a previous existence at a different type of work. One of the last to be seen on the Birmingham system started its working life pulling a milk float. Perhaps the most famous was Bonnie, a former circus performer that could still do some of the routines she had used in the ring. Where the water was not polluted, the horses used that as their drink supply, but in heavily industrialised areas such as the Black Country, they had to have alternative supplies.

The horse was one of the boat family's most prized assets and had to be well looked after. This horse is having a feed from a nosebag and has a crocheted bonnet to protect it from the sun.

The boat families not only ran the boats, they also did the loading and unloading themselves on most occasions – and when it came to loading, they much preferred it that way. They knew how their particular boat behaved and how to distribute the load for maximum efficiency to make sure the vessel 'swam', moved smoothly through the water. It could be hard work. Coal had always been one of the most important cargoes on the canal, and unloading a boat just got tougher as the work went on. At first, shovelling the coal out onto the wharf was simple, but as they got nearer the bottom of the boat, they had to throw the shovelfuls ever higher to get them onto the wharf. Canal work was not romantic, but those who had left it often looked back on those times with nostalgia. Nell Cartwright, interviewed by the Mikron Theatre Company, for their excellent show about those times *I'd Go Back Tomorrow* in 1977, expressed the views of many:

> I would do it all again exactly the same as I had with the horses, the boats, the loading – I have loaded and emptied 25 tons of corned beef, I have emptied 31 tons of spelter, I have done 25 tons of timbers – to me work was nothing. I couldn't care less, I don't even today.
>
> But I liked it and I liked my horse and I liked the boat as it was … I mean you look along the boat as it was going and you see that horse just walking along that road and the hedges and the trees and everything going by. No one could ask for better than that. What I would like to see now – before I leave this world – I would like to see all those horses come back and the place come back as it was as I knew it.

The narrow boat was by no means the only craft to be seen on Britain's canals. Brindley may have been the man of the moment as far as the early promoters in England were concerned, but other engineers had very different ideas. The Leeds & Liverpool Canal was the first to be built across the Pennines. Work started in 1770 – though it wasn't opened throughout until 1816. At the Leeds end, which was completed as far as Skipton by the early 1770s, the canal at first closely followed the line of the River Aire, but left the valley by a series of ever greater leaps. Where Brindley had always avoided obstacles, the engineer John Longbottom charged straight on, culminating in the great five-lock staircase at Bingley, carrying the canal some sixty feet up to a long pound, lock free for seventeen miles. This was canal building on the grand scale, but there were disputes. The company had two committees, one in Lancashire and one in Yorkshire, and they failed to agree on one of the most fundamental questions – how big should the locks be? Both ends had locks that could take vessels up to 14ft 3in wide, but from Liverpool to Wigan, they could take vessels up to 72ft and from Wigan to Leeds just 62ft. As the system developed, it meant that narrow boats from other canals would be able to join the Leeds & Liverpool and head for Liverpool, but not to Leeds. As there was a clear need to have boats that could cross the whole canal from end to end, the result was what became known as a Leeds & Liverpool short boat.

The boats bore a fairly close resemblance to the barges, working on the Aire and the Mersey. They were generally built with rounded bows and a flat, transom stern. The hull was a carvel construction, with keel and keelson and regularly spaced ribs. Unlike the narrow boat, the crew cabin was below the after

Leeds & Liverpool short boats. Narrow boats could use the western end of the canal as far as Wigan, but on the rest of the canal to Leeds the locks would only take vessels up to 62ft, though they were broad beamed.

Wide boats on the Rochdale Canal in winter. The canal has frozen and they are waiting for the ice to thaw before they can move on.

deck and, of course, considerably more spacious than a narrow boat cabin. The steerer stood on the deck above the cabin, instead of in a cockpit. There was also a smaller forward cabin, usually kept for storing gear, but could be used on a family boat for the children to sleep. Decoration was generally far simpler than on a narrow boat, with coloured geometric patterns on the gunwales and more elaborate scrolls on the stern. The same boats were also used on the Lancaster Canal.

Two other trans-Pennine canals were authorised in 1794. The Rochdale was built to take boats that were the same dimension as the longer Leeds & Liverpool, while the Huddersfield Narrow was, as its name suggests, exclusively used by narrow boats which, given that it was to contain Britain's longest canal

tunnel, is not surprising. Britain continued to have something of a hotch-potch of canals. Some were broad, but were adapted to take narrow boats, though in the case of the Kennet & Avon, the dimensions seem slightly arbitrary. Boats up to 73ft long could travel all the way from the Avon at Bristol to the Thames, but the maximum width varied: 16ft on the Avon river navigation; 13ft 10in on the canal; and 14ft on the Kennet.

The Scottish canals had no direct links to the English system, so engineers were free to choose whatever gauge they fancied. The Forth & Clyde received its Act early on in the canal age, in 1768. Inevitably, Brindley was involved, but the main work was handed to the great civil engineer, John Smeaton. Brindley as always wanted to follow his usual methods, but Smeaton would have none of it. They clashed right at the beginning, while the route was being surveyed, in October 1768:

> Mr. Brindley recommends to begin at the point of partition, because, he says, it is his 'constant' practice to do so, and, in the present undertaking, it seems particularly advisable 'on many accounts': but pray, Mr. Brindley, is there no way to do a thing right but the way you do? I wish you had been a little more explicit on the many accounts. I think you only mention one, and that is to give more time to examine the two ends: but pray, Mr. Brindley, if you were in a hurry, and the weather happened to be bad, so that you could not satisfy yourself concerning them, are the works to be immediately stopped when you blow the whistle, till you can come again, and make a more mature examination?

After that very little was heard from James Brindley. The canal was very different from the ones we have seen so far in that it was designed to take sea-going vessels, 68ft 6in by 19ft 8in by 8ft 6in draught. There were no fixed bridges, so tall-masted vessels could use the waterway to link the west coast at Bowling Harbour on the Clyde to the east at Grangemouth on the Firth of Forth. Later, this canal was linked to Edinburgh through the Union Canal. There was never any question of vessels using sails on the Forth & Clyde as there are 39 locks to negotiate in just over 35 miles. However, there was a busy barge traffic.

There were 'scows', 60ft long and 12 to 13ft beam, able to carry up to 80 tons. Because of the short journeys there was no need for crews to sleep aboard. They were very basic, double ended with a small decked area at each end of the central hold, which was often covered by hatches.

Scotland has two other active canals, the Crinan and the Caledonian. Both were built as ship canals, the first to allow vessels to take a short cut across the neck of the Mull of Kintyre, and the latter to allow vessels to go from coast to coast, instead of round the north coast. There was never a need to design special craft to use these waterways.

As the canal system developed, industrialists began to look at ways in which they could take advantage of water transport where the problems of canal construction seemed either impossible or ruinously expensive. William Reynolds, an iron master from Shropshire, needed access from his works to the River Severn, but the height differences were very great, which

A barge crossing the aqueduct that carries the Glasgow & Edinburgh Union Canal across the River Almond. The horses in the photo are nothing to do with the barge – they are pulling a plough in a canalside field.

would have made the scheme impractical for a conventional canal. But, on the other hand, he was not interested in opening up a route for general traffic, so he opted for a canal that would meet his specific needs. He was involved in two linked waterways, the Ketley and Shropshire canals. Instead of locks, he developed inclined planes and the boats were very basic tub boats – basically square-sided boxes. The first to be built, the Ketley plane, had a double railed track which carried a pair of frames, or wheeled platforms. The wheels at one end were small, and those at the lower end larger, so that the platform always remained level. The tub boats would be hauled along, usually as

Barges on the Union Canal in Edinburgh. The high tiller is a feature of these vessels.

a small train, linked together by chains. At the top of the incline was a lock, into which water was pumped by a small steam engine and one of the tubs would be floated in. As the water was let out, the tub boat settled down onto the frame. At the same time, a second boat at the foot of the plane would be floated onto the lower carriage. The two were connected by cables wound round a drum at the upper level. The boat at the top would be loaded, the other empty. As the brake was released on the drum, the heavier tub began its journey down the slope, pulling the other up on the adjoining track.

One section of the Shropshire Canal runs through what is now the Blists Hill site, part of the Ironbridge Museum complex. It is very similar to the Ketley incline. Altogether the incline is 1,000ft long with a vertical drop of 207ft down to the lower level at Coalport. A similar tub boat canal was built in Cornwall, with a sea lock at Bude and then proceeded inland for 36 miles, with a total of six inclines, traces of which can still be found. Unlike the

The Trench inclined plane on the Shropshire Canal, c.1900. The tub boats are floated onto wheeled cradles and moved via a steam engine at the top of the tracks. The boats move alternately, the one going down balancing the one going up.

Shropshire inclines, many of them used water wheels to power the inclines, since the loaded traffic, mainly of sand for farms along the way, was uphill.

A major change in boat construction came at the end of the eighteenth century. I spent my early years near to a local attraction, Mother Shipton's cave by the River Nidd in Knaresborough. In the sixteenth century she was famous for her prophecies, among which was this saying, 'In water, iron then shall float, as easy as a wooden boat'. This was generally considered absurd. Throw a piece of wood into the water and it floats, chuck in a lump of iron and it sinks. The Shropshire iron master John Wilkinson, however, recognised that the air within an iron hull would provide all the buoyancy needed to keep it above water. In July 1787, his iron barge, appropriately named *The Trial*, was launched into the Severn near the famous iron bridge. 'It answers all my expectations', he wrote, 'and it has convinced the unbelievers who were 999 in a thousand.' The idea was not taken up immediately, but over the years, more and more craft of the waterways were built with iron or later steel hulls.

So far, we have only been looking at cargo vessels, but the canals needed constant attention to keep them open – lock gates needed replacing from time to time, banks needed to be piled, bridges repaired and so on. Canals would have their own maintenance vessels, often very basic, little more than flat bottomed, box-like structures, designed to carry the basic equipment to where it was needed. Some were more complex. Pile driving, instead of being done simply by hand, could use steam power, with the engine mounted on the boat and covered by a basic wooden shelter, as seen in the picture (page 70). Dredging was originally carried out using a wooden boat fitted with a crane, attached to which was a spoon-like scoop. This was pulled along the bed of the canal, then lifted and the mud emptied out. As with piling, this was mostly superseded by steam-powered dredgers.

Winter could bring a special problem – ice. This could bring the whole system to a halt. Canal companies invested in special ice-breaking boats. They had metal sheathing and heavily reinforced bows. A rail ran down the middle of the boat, and

Maintenance boats were necessary to keep the canal in good condition. They were basic craft that were towed to where they were needed. Here the boat is being used for pile driving to strengthen the canal bank.

a number of men stood at each side of the rail, holding onto it. As the boat was brought forward by the horse, they would start rocking it from side to side. The bows would rise up and fall on the ice, smashing it. It was effective up to a point, but in a severe winter, boats could be held up for weeks if the ice was too hard to break. In 1895, a severe winter shut many canals for weeks on end and the boatmen were reduced to collecting money in the streets of Birmingham, with a small boat on a cart and a notice saying 'Frozenout Boatmen!'

It would be possible to fill a whole book by simply going over the many small differences that separated one class of canal boat from another, but hopefully this chapter has given the reader a

Ice was a great menace to the canals, so special ice breaking boats were employed. They had hulls reinforced with iron and were hauled up to crash down on the ice and then rocked backwards and forwards to break it up. This wintry scene is on the Sheffield & South Yorkshire.

good idea of the great variety of craft that used the canals at the time, when the horse was the only available source of power. But a new power source had been available, even before the start of the canal age – the steam engine. The impact of this new power source will be the subject of the next chapter.

CHAPTER FIVE

Steam Power

The first really successful industrial steam engine came into use at the beginning of the eighteenth century. It was designed to meet a very specific need. Coal and metal mines were having to go ever deeper underground, but to do so they had to lift out water from greater depths. Pump rods have to go up and down to force the water to the surface – think of the simplest device, the village pump with its handle. A more sophisticated system could use a water wheel for power, but its use was limited. Thomas Newcomen came up with a solution. The pump rods going down the shaft were suspended from one end of an overhead beam, pivoted at its centre. Gravity would automatically pull the rods down, so all he needed was a mechanism to pull them up. Beneath the other end of the beam, he placed an open-topped cylinder inside which a piston could move, hung from the beam by a chain. He now added a boiler and allowed steam to be forced in under the piston. The cylinder was then sprayed with cold water causing a partial vacuum and air pressure forced the piston down, lifting the rods at the opposite end of the beam. Pressure equalised, gravity again took over and so the process was endlessly repeated, the great beam nodding to and fro like a seesaw. It worked, but was a huge consumer of fuel. This was not a problem at collieries, where getting coal was no problem, but for the metal mines of the south west it represented a huge, but unavoidable, cost.

There was little change for over half a century. Then a young Scots instrument maker, James Watt, at the University of Glasgow, was sent a model Newcomen engine which simply didn't perform. In a moment of inspiration, he spotted the flaw in the design – not, in spite of popular mythology by watching his mother's kettle boil, but by appreciating a scientific fact. The energy was being wasted by constantly having to reheat the cylinder after each dowsing with cold water. His solution

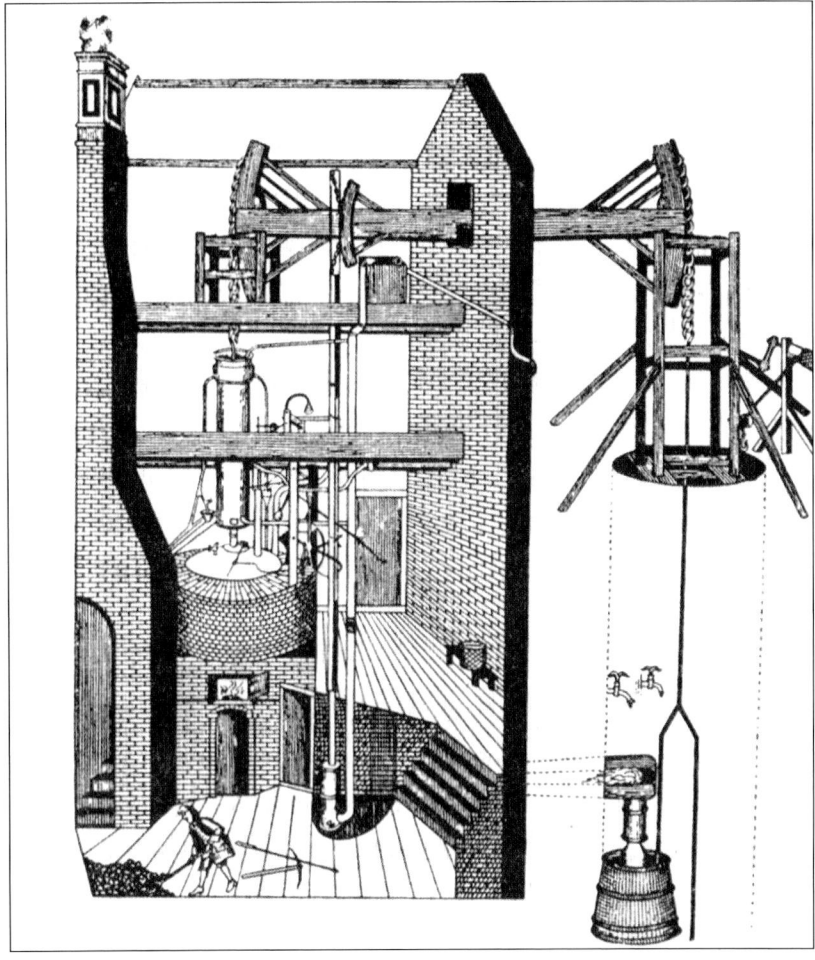

The original Newcomen steam engine installed at Dudley in 1712. The cut-away illustration of the engine house shows the boiler with the cylinder above, attached to one end of the overhead beam, while pump rods attached to the other end that drops down into the mine. The beam oscillates, moving the rods up and down. It was the first successful steam engine.

was that instead of cooling the steam in the piston, it would be cooled in a vessel set some distance away – a separate condenser. However, there was still a problem with heat escaping from the open top of the cylinder. If he closed that off, then air pressure couldn't act on the piston. So, instead of air pressure, he introduced steam under pressure above the piston – the atmospheric engine had become a genuine steam engine.

The Watt engine was a huge advance, but its use was still limited. It was fine for mine pumps, but there was no way the upward stroke of the piston could be used to do any work, when it was suspended from a chain; you can't push a chain up. Simply replacing it by a rigid connection was not the answer, because the end of the beam traced a curved path. This time Watt's solution was one he considered to be his greatest invention. Instead of connecting the rod from the cylinder to the beam itself, he attached it to one end of a sliding parallelogram of rods – the parallel linkage. Now the piston could work in both directions, and now instead of just pump rods a device such as a sweep arm could be fastened to the beam and used to turn a wheel. We have finally reached a point where steam power could be applied to a boat.

The obvious way to use this new form of power on a vessel was via a paddle wheel. For generations, water had been used to turn wheels. All that was needed now was to reverse the usual process. If the paddles turned in water, forcing it back from the boat, then Newton's law of physics, that every action has an equal and opposite reaction, would ensure that the boat would move forward. All that was needed now was someone to recognise the possibility and build a working paddle steamer. One would expect that the answer would have been found first in Britain, then in the middle of a great industrial revolution and the land where steam power had first been developed. It might also seem probable that the inventor would be a British engineer, a man of wide practical experience. In the event, the first successful steamer was built in France by a nobleman, the Marquis Jouffroy d'Abbans. He designed an experimental steamship, the *Palimpède*, in which the engine powered rotating oars. The arrangement was far from perfect, so his next vessel was the paddle steamer, the *Pyroscaphe*. He had hoped to work it on the Seine in Paris, but he was refused permission so instead set up a service on the Saône near Lyon in 1783. He could have been famous as the great innovator, but France was about to suffer the convulsions of a political revolution, and the steamer service was one of the victims. The Marquis, unlike many other aristocrats, did manage to survive. He did write two papers on steam propulsion for boats, but it is doubtful if anyone in Britain either heard of his experiments or read his papers. The next experiment with steam took place in America, where a boat was built which used a steam pump to pull in water at the bows and throw it out at the stern, but the technology was just not available to give enough power for it to work efficiently. So, at last, British engineers took up the challenge, starting again from scratch.

William Symington was born at Leadhills in Scotland in 1764, a small town which was as the name suggests at the heart of a lead mining area. His father worked as a mechanic, but had other plans for his son, hoping he would enter the ministry. He turned out, however, to be far more interested in engineering than theology. His first effort was to design a new type of steam engine which combined elements of the atmospheric engine of Newcomen with Watt's pure steam engine. In the 1780s, he met Patrick Miller of Dalswinton, near Dumfries. He had been experimenting with double hulled boats, with a paddle wheel in between, worked by a crank – rather like the little pleasure boats where the paddles are worked by foot pedals. Miller suggested that Symington should design a steam engine to work the crank, and the result was a small paddle steamer that was successfully run on Dalswinton Loch in 1788. It aroused a good deal of interest, and one of the first passengers was Robert Burns, though sadly he seems not to have written an account of his voyage.

Symington then built a second, larger vessel with a bigger engine, but that failed, largely because the paddle wheels broke up. The experiments did, however, attract the attention of Lord Dundas, who had extensive commercial interests in Scotland and was on the board of the Forth & Clyde Canal Company. He encouraged the engineer to design a steam tug for use on the canal. He tried several experimental models, but the final successful result was the *Charlotte Dundas*. The vessel had steam supplied by a simple waggon boiler to a 22-inch (56cm) diameter horizontal cylinder. The mechanism was very basic. A connecting rod from the piston activated a crank, which worked the paddle wheel set between the hulls at the rear of the vessel. The engine had a comparatively modest rating at 10 horse power, but nevertheless the tug was able to haul two 70 ton barges up the Forth & Clyde against a strong headwind, covering twenty miles in just six hours. The experiment was a success, but the canal authorities were concerned that the paddle wheel churning up the water

William Symington built four experimental paddle steamers for Lord Dundas, intended for use on the Forth & Clyde Canal. One of these, the *Lord Dundas*, was a passenger steamer and had a horizontal steam engine driving a paddle wheel set amidships.

would cause damage to the canal banks, so they refused to allow it to go into operation. However, one canal owner at least was impressed. The Duke of Bridgewater ordered tugs on the *Charlotte Dundas* model for use on his canal. Once again, however, Symington was to be thwarted. Before the order could be put in hand, the Duke collapsed and died. The new owners cancelled the order. What might have been a revolution in canal transport, at least on the broader waterways, came to an abrupt halt. It was to be some time before steam again appeared on the canals.

The next stage of development took place across the Atlantic, where the American engineer Robert Fulton began a regular service between New York and Albany in 1807 with his paddle steamer *Clermont*. There was an important British connection, as the engine was supplied by Boulton & Watt. News of the American service reached Henry Bell, a hotel proprietor from

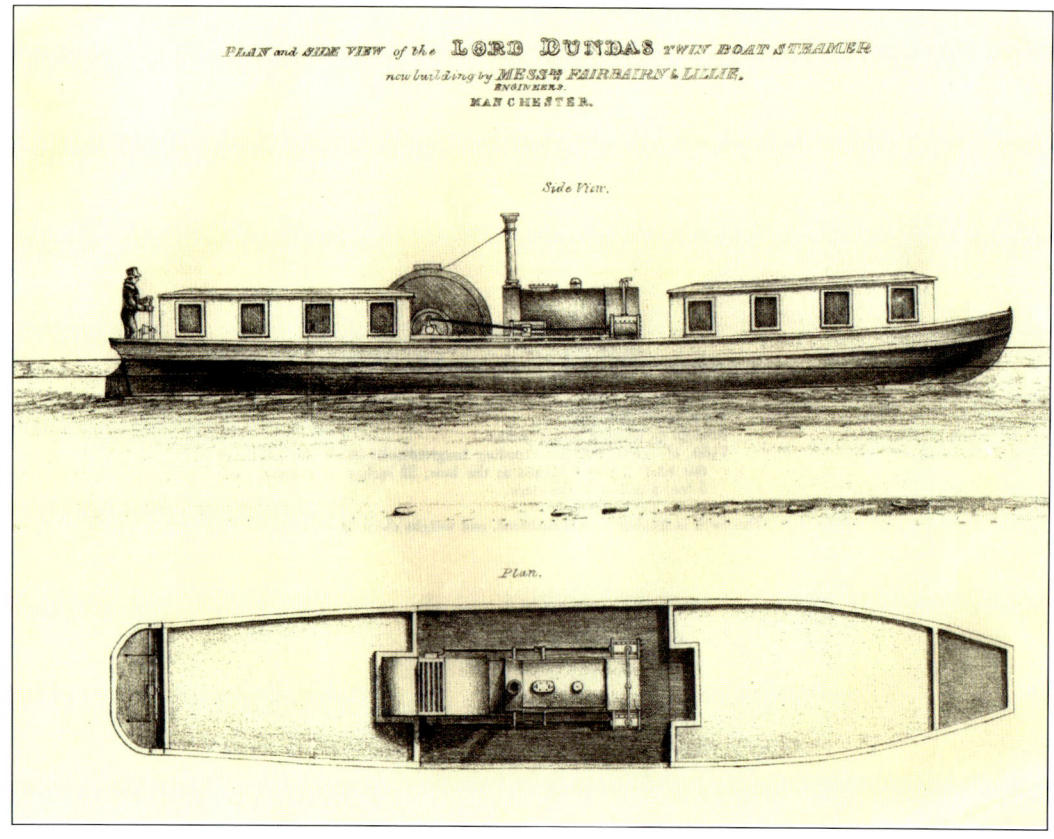

Helensburgh near the mouth of the Clyde, and he decided that he could run a similar service on that river. He had been a regular visitor at a Glasgow forge and foundry, run by a young man called David Napier and offered him the opportunity to manufacture the engine and boiler.

Napier had received an academic education, learning French, Latin and mathematics, but from the age of around twelve spent more time at the family works than in the classroom. By the time he was twenty, he had mastered all aspects of the business and when his father died, he took over. He and his wife had a grand total of fifteen children, which Napier seemed unwilling to distinguish between, usually referring to them as 'the swarm'. The engine he constructed was more complex than Symington's. The 12½ inch diameter cylinder was vertical, with the drive transmitted through side levers to the crankshaft, which had a 6ft diameter flywheel. Again, it was not especially powerful, rated at just 4 horse power and a sail was also used, which in the illustration (page 79) appears suspended from a yard attached to

The best known of the Symington paddle steamers is the *Charlotte Dundas*, the fourth to be built and intended for use as a tug. The main change in design was moving the paddle wheel to the stern.

the funnel. Not having experience of this sort of work, Napier had quite a few problems to solve:

> I recollect that we had considerable difficulty with the boiler, not being accustomed to make boilers with internal flues, we made them first of cast iron but finding that would not do we tried our hand with malleable iron and ultimately succeeded with the aid of a liberal supply of horse dung in getting the boiler filled.

The vessel was named *Comet* and worked at first on the Clyde and later further up the west coast and in 1820 ran aground at Creignish Point on the Sound of Jura. The service had proved very popular, though not all approved. A clergyman from Crinan denounced Bell and his craft:

> He, being puffed up in pride of his abilities, did lately conceive a ship to go upon the waters, not by dint of the clean winds of the air as ordained by God, but by means of fire with a great smoke which issues from the bowels thereof.

One man who did not profit directly from the success was David Napier as it seems that Bell never got round to paying him. But he did go on to make a highly successful career as a shipbuilder. And the little *Comet* began a long tradition of Clyde steamers, taking Glaswegians 'doon the water'. The last of these, the *Waverley*, still operates, but is a far grander and more powerful craft than Napier's vessel. Pleasure steamers became a regular feature on many of Britain's rivers and the last paddle steamer to be built for river excursions '*Kingswear Castle*' still operates on the Dart as she did when brand new.

The steamer was built in 1924 at Falmouth, but the engine came from an earlier vessel of the same name, built twenty years earlier. The progress made in the century since *Comet* took to the water is obvious. She is over 100ft long, able to carry 235 passengers. The earlier craft had a simple single-cylinder engine, but here the engine is a compound. Over the years, steam pressure was gradually increased to a point where the exhaust steam was still under pressure. Rather than

The paddle steamer *Comet* of 1812 was the first British steamer to go into commercial service, operating for excursions from Helensburgh on the Clyde. The tall chimney also acted as a mast for a square sail.

let it simply puff away into the atmosphere, it is here fed into a second cylinder. In order for the two to work in balance, the second cylinder has to be larger than the first. In this case, the high pressure cylinder is 12in diameter, the low pressure 24in. I have a family connection with this vessel. A few years ago, a new boiler was installed, but before that it had been supplied by Riley Brothers of Stockton-on-Tees and one of the three brothers was my great grandfather. Paddle steamers have provided pleasure for vast numbers of holiday makers, who enjoy the delights of some of our most picturesque rivers, but they were also to find a more purely commercial use on inland waterways.

There was one obvious area for development. The cargoes of the world were still being carried by sailing vessels, but there were often problems in the final stages of their voyages, when they had to travel up river to berth. Major ports, such as London, Bristol and Newcastle, were some distance from the open sea and an unfavourable wind could leave them either unable to get into harbour or stuck in and unable to leave. Marc Brunel was one of those who had been impressed by the commercial possibilities of steam excursions and had converted an old sailing vessel into a paddle steamer, which ran excursions from London, round the

Kingswear Castle is the last surviving coal fired paddle steamer in the world. Built in 1924, she runs excursions on the River Dart as she did a century ago.

coast to Margate. He had already had considerable dealings with the Admiralty and suggested to them that paddle tugs would be immensely useful to the Royal Navy. He had not allowed for the deeply ingrained conservatism of that august body and received this rather astonishing reply. 'Their lordships feel it is their bounden duty to discourage the employment of steam vessels as they consider the introduction of steam a fatal blow at the naval superiority of the Empire'. Quite why towing a man of war into port would threaten the empire was not made clear, and there is something Canute-like in supposing that because the Admiralty disapproved of steam, the rest of the world would agree with them. Commercial concerns, however, took a different view, and paddle tugs soon appeared as important aids on major rivers. The most famous depiction of a paddle tug at work is Turner's painting *The Fighting Temeraire* where the veteran of Trafalgar is being towed to the breakers' yard. It could also be taken as a

metaphor of the old world of sail giving way to the new age of steam, whether the Admiralty liked it or not.

A major change in the development of steam on the water came thanks to a farmer, Francis Pettit Smith. He was looking for an alternative to the paddle wheel and turned to a device first developed by the famous Greek mathematician, Archimedes, in the third century BCE, the Archimedean screw. This consisted of a helical screw inside a cylinder. It has been used for centuries for drainage. If one end of the screw is immersed in water and the screw rotated either by hand or by a mechanical device, then the water will be raised up the cylinder. He reasoned that just as the paddle wheel pushed water backwards and moved a boat forward, so a screw could have the same effect. His first experiments were with models on the duck pond at his farm.

The engine of *Kingswear Castle*. Unlike the simple engine on *Comet*, this is a two-cylinder compound. The steam that leaves the first cylinder is still under pressure so, instead of being allowed to exhaust it is fed into a second, low pressure cylinder.

One of the important uses found for paddle tugs was towing sailing ships into and out of harbours situated up rivers. This tug is hauling a vessel to the Pool of London from the mouth of the Thames in the late nineteenth century.

He started with a screw consisting of two full spirals, which seemed logical – the more the better. Then one of the screws broke in half and proved to be more, not less, efficient. He was now ready to go to a full-scale trial with a launch, the *Francis Smith*. In 1836 he took out a patent for his invention, the screw propeller, and set up the Propeller Steamship Company. He received financial backing to build a steamer, the *Archimedes*, which was sent round Britain to promote his invention. One man who was impressed was Isambard Kingdom Brunel, who changed the design of his iron-hulled ship, the *Great Britain*, scrapping the planned paddle wheels in favour of a propeller.

The arrival of the propeller had an impact on all forms of water transport, including the craft of the inland waterways. Symington's tug for the Forth & Clyde had been rejected because of the turbulence from the paddle wheels. The propeller did not have such a violent effect, and steam tugs found a wide variety of uses. Long tunnels could mean long delays, as boats were slowly legged through. With a tug, a whole string of boats could be hauled through at a time. One set would be pulled through, while the horses were led over the top to wait at the other end. Then boats travelling in the opposite direction could be brought along on the return trip. This became a regular feature on the Grand Junction Canal, for example, for as long as the horse boats dominated the system.

STEAM POWER • 83

Tugs were invaluable for towing narrow boats through tunnels. The Braunston tunnel on the Grand Union is over a mile long and otherwise boats would have to be legged through. Here the boats line up to be pulled in a string by the steam tug on the left. The horses would be led over the top.

Crews and horses waiting as a tug emerges from the tunnel with the narrow boats in tow.

Some companies realised that the tugs could be used to replace horses on the waterways. The Severn and Canal Carrying Company had a regular trade supplying the Cadbury factory at Bournville. Cocoa beans were imported and brought up the Severn to Sharpness, where they were loaded directly into narrow boats. A whole fleet of these, perhaps as many as twenty boats, would then be taken up the river to Worcester. The narrow boats were specially adapted to cope with the river, usually with timber heads on each side of the deck, with holes drilled through. The tow ropes were then passed down each side and down to the next boat in the line. This gave added stability. At Worcester, the boats continued up the Worcester & Birmingham, reverting back to horses to pull them along. However, on the long pound between Tardebigge and Birmingham, steam tugs were again used for haulage through the tunnels. On the return journey, the hold had to be thoroughly scrubbed out.

These were by no means the only craft to take advantage of steam tugs. As mentioned earlier, sailing barges such as the Humber keels often had to rely on horses when negotiating canals such as the Sheffield and South Yorkshire, so this was a job that could easily be taken over by steam tugs.

The Severn & Canal Carrying Company used steam tugs to tow narrow boats on the Gloucester & Sharpness. The tug is seen here leaving Gloucester Docks.

If the propeller could be used on canals, why should it not be applied to the narrow boats themselves? Steam narrow boats were, in fact, introduced onto the Grand Junction Canal in the 1860s, but they had their own problems. Firstly, a great deal of space had to be taken up to provide an engine room and coal bunkers. Then they needed an extra man, who understood steam engines, to work down in the engine room. On the other hand, the steam narrow boat could act as a tug, pulling a conventional boat behind it. For carrying companies, it was a question of balancing the gains in efficiency with the extra costs. One company that built up a fleet of steamers was Fellows, Morton & Clayton. They employed all male crews, and worked the boats fly, constantly on the move. Few other companies followed their lead. The crew were distinguished by their white uniforms, not perhaps ideal for shovelling coal into a boiler in a confined space.

Steam tugs were also used on the waterways of the north east and are here seen with a pair of Humber keels in tow on the River Trent.

The invention of the screw propeller to replace the paddle wheel made it practical to install steam engines in narrow boats. Fellows, Morton & Clayton were one of the few companies that used steamers. The disadvantage can be seen in the photograph; cargo space is lost to the engine room forward of the rear cabin. However, the steamer could also be used to tow a former horse boat behind.

If there were problems in using steam power in narrow boats, the same did not necessarily apply on wider canals. A steamer was specially designed for use on the Forth & Clyde Canal. In use, the exhaust steam was allowed to escape up the funnel, resulting in regular puffs of smoke at each stroke of the engine, and the craft soon became known as Clyde Puffers. They were originally simple adaptations of the existing barges on the canal, the sailing gabbarts, that were roughly 70ft long by 18ft wide. Over the years, several changes were made. Originally, steering was by tiller from an open position at the stern. However, the steerer found it difficult to see over the boiler, so a built up section was added, later given some protection from the elements by a canvas dodger. Eventually, a proper wheelhouse was provided with wheel instead of tiller steering. The puffers proved to be very handy craft and were soon finding more extended use, travelling up and down the west coast of Scotland, serving mainland ports and islands. They were particularly useful for island communities that had no piers as the puffers had almost flat bottoms and could be

beached at low tide for loading and unloading and floated off again at high tide. But going to sea presented a new problem. On the canal, the boiler could be refilled with canal water, but sea waters and boilers are not a good combination. The answer was to add a condenser. Now the steam, instead of going up the funnel, was condensed back to pure water – but the puffers no longer puffed, simply emitted a steady stream of smoke from the funnel. During the Second World War, the Admiralty needed a craft to service the fleet with supplies and had no need to find a new design. They began ordering puffers, but gave them a new name. They became victualling inshore craft with the initials VIC for identification. One of these, VIC 32, is one of the very last puffers still in steam.

The VIC 32 was built in 1943 and was eventually bought by Nick and Rachel Walker in 1975, who set about converting her for passenger carrying. Outwardly, she remains much as she always was, though the hatches have been slightly raised. There is still the captain's cabin in the stern and foc's'le for the crew.

One very successful canal steamer was the Clyde Puffer, designed for use on the Forth & Clyde. The name originated from the fact that exhaust steam was puffed up the funnel. One of these sturdy vessels is here seen loading barrels of whisky at Port Glasgow.

In between, however, the hold has been divided horizontally to create a saloon and galley area, with six cabins down below. One area that has never changed fundamentally is the engine room. There is a two-cylinder compound engine and a vertical boiler, with coal bunkers to either side – when loading with coal, there is the rather dirty business of moving half from the bunker beside the quay to the opposite side to keep the vessel in trim. There is a steam winch in the bows.

I took my first week's holiday on board some forty years ago and have made many trips since – and shovelled a good few tons of coal into the boiler. I should perhaps make it clear that helping out on board is not compulsory in any way – but many passengers really welcome the chance to get involved in the working life of the boat. Communication between the engine room and the wheelhouse is by speaking tube and on one occasion we were travelling down the west coast of Scotland and approaching a measured mile. These are landmarks deliberately placed so that passing ships can get an accurate measure of their speed. The VIC 32 normally maintains a steady, if unspectacular, 8 knots. Nick called down to me that he would be quite keen to see just how fast she would go. I set about raking over the fire, spreading coal and getting a good steady heat. The indicator on the pressure gauge crept up to the red mark, indicating the maximum pressure available at 120psi, beyond which the safety valve would start blowing off. I managed to keep just below the mark for the whole mile – and we still sauntered along at 8 knots. Puffers, it seems, are not to be hurried.

The Clyde puffer is not an elegant craft, but a sturdy, no-nonsense working ship, and none the worse for that. Yet puffers seem to arouse a feeling of affection in all who come across them. Perhaps, it is because, they are so basic, rather like a toy boat a child might have in the bath. For some, however, there is a more specific reason – they bring to mind the Para Handy stories by Neil Munro. Para Handy is skipper of the puffer *Vital Spark* which in his own eyes, if no one else's, is 'fine enough to be a yacht'. He and his crew are unreliable, constantly in trouble with the authorities, and if you had to deal with them in real life on a business footing, they would drive you to distraction. But

The preserved steam powered Puffer VIC 32 on the Crinan Canal. Originally built in the 1940s for the Admiralty as a Victual Inshore Coaster to supply the fleet, she now carries passengers on the Caledonian Canal and around the west coast of Scotland.

for the reader, they are simply characters who bring a smile to the face and possibly a dram to the lips as a salute.

Steam tugs had been in use on the Aire & Calder since 1831, but as part of a major improvement system for the waterway, the engineer William Bartholomew introduced a new type of compartment boat to the system. These were known as Tom Puddings, simply because the basic iron tub boats were roughly the shape of pudding tins. Trains of up to fifteen boats, or 'pans' as they were often called, could be taken out at any time, but the one nearest the tug had a false bow. This was also known as a 'jebus' and was used to prevent the water thrust back by the propeller building up in front of the compartment boat. The Tom Puddings were specifically designed to take coal from the mines of South Yorkshire to the port of Goole, for exporting round the

Compartment boats known as 'Tom Puddings' were developed for use on the Aire & Calder in the late nineteenth century. A train of these vessels would be loaded, as here, with coal and taken by tug to Goole where they would be unloaded into coastal colliers.

coast. They would be loaded at the colliery in the usual way from staithes that dropped the coal into the pans. The unloading at Goole used special hoists. A pan would be detached from the train and floated in under the hoist. There it would be picked up, carried to the top of the hoist, then upended and the coal discharged into a chute that took it directly into the hold of the waiting ship. At the beginning of the twentieth century, there were estimated to be around 1,000 Tom Puddings at work.

I was travelling down the Aire & Calder in the 1970s, when Tom Puddings were still at work, though no longer hauled by steam. A lock keeper told us to expect to meet them somewhere along the way and offered this advice. 'They'll not stop for you. They'll have you up the bank as soon as look at you.' It is only courtesy for a pleasure boat to give way to a working boat, but one look at the Tom Puddings was enough to make sure we took the advice. It did give me an opportunity to see how they coped with going down the lock. There were fifteen pans in the train, far longer than would fit into the lock. I was curious to

Tom Puddings at Goole. They are being floated under the hoist, which lifts them to the top and then upends them to send the coal down the chute into the hold of the waiting ship.

see how this would be managed. First the hawsers were hauled in, pulling the dummy bows right up to the stern of the tug. Then the train was broken in half and the tug and seven pans went down the lock, leaving the rest behind. Once the first set had left the lock, it was refilled and the top gates opened. Then the bottom paddles were opened and the flow of water simply sucked the remaining pans in. The top gates were closed, and when the levels were down, the train was reunited to continue its journey to Goole. There are no longer any Tom Puddings working, but their story is told in a museum at Goole.

Although steam was not readily adapted for use on narrow boats, there was no reason not to use it on wider waterways. Among the barges that were transformed were the Weaver flats.

Weaver steam flats on the river at Runcorn. The vessel in midstream is one of a fleet run by ICI.

Steam found other uses on the canal system. During construction of the Caledonian Canal, steam dredgers were extensively used, and dredgers were later used to keep canals clear of silt. One canal that made particular use of these machines was the Gloucester & Sharpness. In 1849, the depth of the canal had been reduced from its original 18ft to just 14ft by accumulation of silt. The canal company ordered their first dredger to deal with the problem. It was successful, and in two years of operation, the machine lifted out 150,000 tons of mud. More dredgers were supplied over the years. Steam dredger Number 4 was built by the De Klop shipyard in the Netherlands. This was a bucket dredger. The buckets, each able to hold 7cwt, scooped up the mud, then as they travelled back up, were emptied into a barge moored alongside. Winches on board allowed the dredger to move sideways, through wire ropes fastened to the bank. If it had to be moved long distances, it had to be towed by a tug. The thoughtful Dutch provided two pairs of clogs when the boat was delivered, one for the fireman and the other for the engine driver, to keep them comfortable when walking around on the hot, oily plates. Number 4 is now preserved as a working exhibit at the waterways museum at Gloucester Docks.

Steam had an important part to play in improving transport on the rivers and canals in the nineteenth century, but most of the system still relied on the sailing barges and horse-drawn craft. That changed totally in the twentieth century.

CHAPTER SIX

Motor Boats

The first proposal for a form of internal combustion engine was made as early as 1680, when the Dutch scientist Christian Huygens suggested exploding gunpowder in a cylinder fitted with a piston. His idea was that the hot gases would be allowed to escape through a valve and then, as the cylinder cooled, a partial vacuum would be formed under the piston. Then, as in the Newcomen engine, air pressure would push the piston down. There was an obvious snag in turning the idea into a working model – how do you safely recharge the cylinder with gunpowder? Huygens abandoned the idea, but his assistant, Denis Papin, went on to develop a rudimentary steam engine. Nothing more was heard of internal combustion for another 200 years. The next more practical idea used a more suitable substance in the cylinder – gas, which was then being made from coal. A patent was taken out for a gas engine in Britain in 1794, but it was to be some time before a really successful engine was developed that could compete with steam power.

Nikolaus Otto, a German scientist, built a horizontal gas engine in 1876. It worked on what became known as the Otto cycle, consisting of four strokes, controlled by valves. In the first stroke, the explosive mixture is drawn into the cylinder; in the second the mixture is compressed by the piston and ignited by an electric spark; in the third, the explosion drives the piston back again; and on the fourth stroke, the gases from the explosion are exhausted. It was an immense success and the firm of Otto and Langen sold around 50,000 engines in just 17 years. The engine was, however, firmly anchored to the spot, relying on the gas supply to be piped to the site. It was possible to create gas by burning coal on the boat, and there were experiments, first begun by the boat builders Thornycroft of Chiswick, who installed a 30hp gas engine in a boat '*Duchess*' in 1906, but few followed his example. To convert this system to one that could be used for transport required a different fuel.

The answer was found with the development of petroleum and fuel oils. The most significant advance was made by Rudolf Diesel. He used the comparatively heavy oil as fuel instead of the lighter petroleum, and built an engine based on the Otto cycle. Once the engine has been heated up, the heat of compression will allow it to keep running. Its success can be gauged from the fact that such engines are still known by Diesel's name and so is the fuel they use. But the man who developed the engine for use in canal boats was a Swedish engineer, Erik August Bolinder. He was born at Stockholm in 1863 and joined the family firm of machine manufacturers, which he took over on the death of his father. In 1893, he designed a paraffin engine, but soon decided to concentrate on heavy-oil engines for marine use, working at much the same time as Diesel. In 1908, he designed a simple, single cylinder two-stroke engine and like all diesel and semi-diesel engines of the time, it had to be preheated before it could be run continuously. In this case, he used a blow lamp to heat the fuel in a vaporising bulb. This must have seemed quite alarming to anyone coming across this type of engine for the first time, especially if the engine was to be installed in a wooden boat. The Bolinder salesmen had a special party piece to reassure potential customers. They would throw some fuel into the bottom of the boat and chuck in a lighted match, which was at once extinguished.

The first Bolinder to arrive in Britain was fitted to a Thames lighter, the *Travers* in 1910, and was an immediate success. In 1911, Cadbury's built a special narrow boat, the *Bournville 1*, and installed a Bolinder. It was the start of a fleet of boats that ended with *Bournville 17* built in 1926. By 1930, there were 230 other narrow boats fitted with similar engines. They had a distinctive 'pop-pop' sound that soon became as familiar on the canals as the clatter of horses' hooves had once been. There was another unique noise – the loud backfire. By pulling a lever on the engine, the boatman could alter the ignition sequence, creating the bang and putting the engine into reverse. It was not quite that straightforward; the boat had to be slowed right down to very low revs, before the backfire would work, bringing the boat

The Bolinder semi-diesel, first developed in Sweden, was the first internal combustion to be used in barges and narrow boats in Britain. This example is on display in the London Canal Museum.

to a complete halt, before setting off in the opposite direction. A Number One who had just had a new Bolinder fitted at a boatyard on the Wendover Arm of the Grand Union had not really taken the instructions on board. He was heading off at full tilt towards the main line and thought all he had to do was pull the lever. But he was mistaken – he had forgotten to slow down first. He finished up with the bows buried deep in the opposite bank. He had to haul the boat back to the yard by hand and get it repaired but he did not make the same mistake again.

The arrival of the Bolinder and other diesel engines that followed over the years made a huge difference to the life of the boating community. For those who owned or worked sailing

This unusual looking narrow boat was built for Cadbury's in 1911 and was the first in Britain to be fitted with a Bolinder engine.

barges, they were no longer held back for lack of wind on the open water, nor did they have to rely on horses to come and move them along canalised sections. The greatest effect of all came with the families working the narrow canals. One of the great problems with the steam narrow boats was the necessity to have an extra hand to look after the engine. With the motors, that was no longer necessary. The steerer had the controls ready at hand. He still had the job of getting it started, by heating the bulb of a Bolinder. Some boatmen were convinced, quite wrongly, that to get the best out of the engine you had to keep going till the fuel overflowed and flames shot up the chimney. Later, other engines, such as the Nationals, were introduced, which used a starting handle instead of a hot bulb. That involved keeping it turning vigorously to create enough heat of compression. It was hard work and not without its dangers. If it backfired you could end up with a strained or even broken wrist.

It was all very well getting an engine started, but to keep it running well required regular maintenance by a skilled mechanic. Some boatmen prided themselves on their abilities. One I met, Alex Purcell, claimed he could sort anyone's engine problems and was prepared to do so for a suitable payment. But he would never let anyone see what he did down in the engine room. Boatmen did not see that as unusual. It has to be remembered that the engine room was also part of the family's floating home. Even on company-owned boats, the official fitters had to ask permission to come on board.

Most readers will be familiar with the arrangements on a motor boat, as they formed the basis for the vast majority of hire boats and privately owned pleasure narrow boats on the canals today. The motor is immediately distinguishable from the horse boat, by its rounded, counter stern, swan neck metal tiller and metal rudder. There were other types of engines used, including the 'Hook detachable Motor'. It sat on the cabin roof and was used to drive an outboard propeller. It was not a success. Steerers found it difficult to see where they were going and found it hard to steer the boats. There were frequently problems with an empty boat, riding high in the water, when the motor would get stuck under a bridge. So the vast majority of companies used diesel or semi-diesel engines, and Fellows, Morton & Clayton, who had been pioneers in the use of steam, once they had acquired Bolinders stayed with them for as long as the company lasted.

The arrival of the diesel engine meant that boats could now work in pairs. The former horse boat would now be the 'butty' towed behind the motor. The tow rope would generally be attached to a stud immediately in front of the butty steerer. This meant he or she could adjust the length of the rope as circumstances changed. Generally, boats worked with a separation of around 60ft. With the rope at hand on the butty, the distance could be altered to take account of sharp bends, especially when travelling on rivers. On the Grand Union, the wide locks allowed two boats to fit in side by side, so it became common to use a short line, attached to a stud in the bows of the butty. Sometimes, especially when running empty, boats were breasted up, running side by side, either fastened together by ropes to studs or to special hooks built into the cabin sides.

Manoeuvring a pair of boats into a double lock was always a slightly tricky business. The motor would enter first and cast off the tow rope, allowing the butty to glide slowly in. If it seemed

The arrival of the motor boat marked the beginning of the working pairs as seen here, passing the maintenance yard at Bulbourne on the Grand Union. The motor boat is towing the unpowered butty.

One of a pair of Samuel Barlow's boats. The captain of the butty, seen here, is able to control the length of the tow rope attaching him to the motorboat.

Sometimes when running empty, a pair would breast up, running side by side. This pair is on the Thames.

to be going too fast, then it could be steered against the side of the motor to slow it down or be stopped by strapping a bollard. On canals with narrow locks, the butty had to be left with its nose up against the lock gates, waiting to be towed in. As a result, working locks with a pair was slower work than with a single boat. But for the boatmen time was important so they had various tricks for speeding things along. Going down a lock, a line could be fastened to the hand rail of the lock gate, and when the levels were about right, the motor would reverse, pulling the gate open. The line was then pulled back in by the butty as the boats left. Going uphill, the motor would rev its engine and the water pushed back by the propeller would help close the gate, and at the same time, the paddles would be raised at the far end to start filling the lock.

Once it was introduced, the motor boat made a huge difference to the families who worked narrow boats. They now had two cabins instead of just one, doubling their living space. Unlike with the steamers, there was no need to employ extra hands, and it was usual for the husband to be in charge of the motor, with the wife following on behind with the butty. This was a time when it was generally believed that women neither understood nor could cope with machinery. That this was simply not true was proved during the Second World War, when women were recruited to run boats, while many boatmen left for the services.

A working pair heading into a lock on the Grand Union, where they will line up side by side.

The all-women crews had to learn quickly, for few if any of them had previous experience of canal boats, but soon proved they were just as efficient as the men.

Living conditions may have been more comfortable, but the work was no easier. Just as crews made going through the locks as efficient as possible, so there was no time to stop for breaks for tea. Instead, a brew would be made up in the cabin of the motor, and a mug of hot tea put down on the towpath under a bridge, where it could be scooped up by the steerer on the butty. Reluctance to stop even extended to the occasional emergency. One boatman told of how one of the children fell off the cabin roof into the canal. He simply shouted the news back to his wife, who hauled the infant out of the cut by his hair on her way past.

Motorised barges became common on Britain's broader waterways. In the north east, several companies developed fleets of steel hulled vessels, some of which were specialised. One valuable trade was in fuel oil, which would originally have simply been carried in barrels on conventional barges, but in the latter part of the twentieth century, special tankers were designed, the largest of which had a capacity of around 600 tons. Motorised barges continued to use the north eastern

The most obvious visual difference between the motor and the butty can be seen at the stern. The motor has a metal, swan neck tiller and the butty a wooden tiller attached to the stern post and decorated with ornate ropework.

waterways, long after narrow boats had stopped. Travelling on the Sheffield & South Yorkshire in the 1970s, I followed a typical barge, in which the open steering position and tiller of the older vessels had been replaced by a substantial wheelhouse. On the same trip, after joining the River Trent, we were overtaken by two coastal vessels that were able to travel inland on the tidal river as far as Gainsborough. To someone on a small narrow boat, they seemed enormous but by coastal shipping standards, they were quite modestly sized. Travelling these waterways was a useful reminder of the rich variety of trade on the tidal rivers and canals of the region. The first port of call for many ships arriving from the North Sea was Hull. From here cargoes could be transhipped into smaller vessels and barges to enter a complex of rivers and canals, stretching far inland.

Just as large ocean going vessels could travel far up the Trent, so too they could use the tidal Severn. Sharpness became an important interchange point between ships and a variety of barges and narrow boats. One of the most important carriers in the region was the Severn and Canal Carrying Co. (SCC). As described in the previous chapter, by the beginning of the nineteenth century they had as many as eighty narrow boats and a number of steam tugs. SCC also had a number of barges, based

A motorised barge with enclosed wheelhouse on the Sheffield and South Yorkshire in 1976.

An array of motor barges in a dock at Hull in 1989.

at Gloucester, mainly for use on the Thames and Severn Canal. They were mostly dumb barges, but they introduced their first steam barge in 1869 and in 1913, they brought in their first two motor barges, *Osric* and *Serlo*. In 1932, the company introduced its own design of motor boats, starting with *Severn Trader*, built in Bristol. She was 89ft long and 19½ft beam, but with a comparatively shallow draft of just 7½ft, when fully loaded with 175 tons. This meant she could travel up the Gloucester and Sharpness Canal and rejoin the river to continue to Stourport. She was too large to enter the basins at the end of the Staffs & Worcester, so could only unload at the quay. A year later, she was joined by *Severn Carrier* to the same design but fitted out as an oil tanker. Other motor barges followed, including

Severn Collier designed as the name suggests for the coal trade, but unusual in having a wooden hull, unlike the steel hulls of the other barges.

The Thames, like the other major rivers, was soon busy with motorised barge traffic. Limehouse Dock was a major transhipment point between the deep-sea ships and canal craft and in its heyday would have been served by a variety of barges, lighters and narrow boats. Barge traffic is a fraction of what it once was since the closure of most of the docks following the development of the container port at Tilbury. However, there was a welcome increase in water transport when barges were used to take construction material up the River Lea during the building of the Olympic Stadium. Most of the barge traffic, though, has sadly diminished, but many of the old vessels have been converted into houseboats and a few have been converted for pleasure boating. A few years ago, I had the pleasure of a week on French waterways aboard the former Severn barge, *Pisgah*.

The Severn & Canal Company used a mixture of vessels, large barges and tugs to work the Severn and narrow boats to carry cargoes further inland.

Sharpness Docks, where the Gloucester & Sharpness Canal joins the Severn, was still being used by barges and small coasters when this photograph was taken at the end of the twentieth century.

Limehouse Dock, at the junction of the Regent's Canal and the Thames, was one of the busiest interchange ports on the system well into the twentieth century. Here cargo from a sea-going ship is being exchanged with a fleet of narrow boats.

Many barges found new uses once their trading days were over. Here, the author enjoys a French waterways holiday on a former Severn barge.

So far, we have been looking almost exclusively at cargo vessels, but from the early days there was a profitable role for specially built passenger boats.

Chapter Seven

Passenger Boats

Navigable rivers were often able to offer a far better service than could coaches on the roads, especially in the years before the Industrial Revolution, when the roads themselves were often impassable in bad weather. Even when the turnpike trusts set about major improvements, the results could be little better than a farm track. This is Arthur Young, describing the Wigan turnpike in 1770.

> To look over a map, and perceive that it is a principal one, not only to some towns, but even whole counties, one would naturally conclude it to be at least decent; but let me most seriously caution all travellers who may accidentally purpose to travel this terrible country, to avoid it as they would the devil; for a thousand to one but they break their necks or their limbs by overthrows or breaking downs. They will meet with ruts which I actually measured four feet deep, and floating with mud only from a wet summer; what therefore must it be like after a winter.

Travelling by river must have seemed a better option when it was available. Not that river travel was without its hazards, especially in the days of the flash locks. In 1634, a passenger boat with 64 on board overturned while riding down the flash at Goring on the Thames. There were no survivors. The arrival of pound locks at least meant that such accidents would not recur. The river was at its busiest in London. For centuries, the only bridge across the river within the city was the old London Bridge, with its narrow passageway, hemmed in by buildings. This meant that far and away the most efficient way of crossing the river from bank to bank was by boat. There was also a regular service going up and down the river. In general, boats that plied for hire worked either exclusively below or above the bridge. Passing through the narrow arches of the bridge was treacherous, especially when a strong tide was running.

There were two categories of open passenger boats, the wherries, rowed by six or eight oars, and the one-man scullers or oars. The watermen, who had the exclusive right to provide the service, would attract customers by shouting 'Oars, oars'. There is a story of a French visitor who paid his sixpence and was disappointed to find that all he got for his money was a trip on the river, not a visit to ladies of pleasure. It was an extremely busy scene. John Stow in his *Survey of London* of 1603 recorded '2000 wherries and other small craft' on the river, while E.W. Bayley, writing in 1796, recorded a staggering 3,419 small craft.

A Thames waterman taking on passengers from the steps at Queenhithe.

There was one special class of passenger boat on the Thames quite unlike any other. Where most of the river barges were strictly functional, the barges of the livery companies, the Lord Mayor and the royal family were altogether more sumptuous. The earliest record dates from 1422, when the Brewers' Company recorded that 'Sir William Walderne was chosen Mayor on St. Edmund's Day, when it was ordered that the Aldermen and Craft should go to Westminster with him to take his charge, in barges without minstrels.' These barges were rowed by teams of watermen and were sumptuous affairs, with a cabin for the dignitaries and everything decorated in elaborate style with a wealth of gilding. By the sixteenth century, Lord Mayor's Day had become an annual event, and the river procession would have been spectacular. A procession was led by the Mayor's own barge, followed by all the different liveried companies in order of precedence, with the Mercers at the head and the Clothworkers bringing up the rear. If the royal barge was taking part, it took precedence, with all the other barges lining the bank as it went past and then falling in behind.

Royal processions became quite frequent under the Tudors. The marriage of Anne Boleyn to Henry VIII was a particularly grand affair, with the new queen resplendent in cloth of gold. In just a few years she would be making a less imposing journey by

The Lord Mayor's Procession passing the Palace of Westminster in 1683, a painting by an anonymous artist. The ornate barges of the Livery Companies are accompanied by a number of skiffs and wherries taking passengers to enjoy the spectacle.

barge to the Tower of London for her execution. Elizabeth I was particularly fond of processions, and when her barge took to the water she expected all the companies to turn out to accompany her. We do have an account of such processions under the Stuarts. Samuel Pepys recorded watching one in 1662, when as well as the fleet of barges, there were two pageants presented on the water 'one of a King and another of a Queen, with her Maydes of Honour sitting at her feet very prettily'. Most occasions would have had music of some sort, but we have no record of what was played until the eighteenth century. On 17 July 1717 – a neatly regular date, 17.7.17 – George III went by royal barge from Whitehall Palace to Chelsea and commissioned music from the most famous composer of the day, George Frederick Handel – his well-known Water Music. Fifty musicians were hired to play on a barge accompanying the king, who was so pleased that he told the musicians to wait at Chelsea and play it all over again for the return journey. The oarsmen who worked these splendid vessels were given their own equally splendid uniforms, consisting of a jacket with the royal insignia, knee breeches, white silk stockings and a cocked hat.

River pageantry became less popular in the nineteenth century, mainly because the Thames itself had become more and more unsavoury as all kinds of refuse, including raw sewage, was dumped in the river. By mid-century, things had got so bad that work in the House of Commons became extremely unpleasant and curtains soaked in vinegar were hung in the chamber in a futile attempt to keep the stench at bay. The good news was that it forced MPs to do something about it, and the result was a proper sewage treatment system designed by Joseph Bazalgette that released the treated effluent far downstream, and incidentally required the building of the Embankment. The splendours of river pageants returned, when a new royal rowing barge was commissioned for Elizabeth II's Diamond Jubilee. Named *Gloriana* she was as grand and ornate as any of her predecessors.

The canal system was built in the first place as a transport route for cargoes of various kinds. The Duke of Bridgewater financed his canal specifically for carrying coal from his mine to Manchester. Yet in 1766, just five years after it was opened, the canal records show that some passengers were also being carried, and the Duke decided it would be worth his while to have two special craft built for this new trade. One was able to carry 80 passengers, the other 120, and both had what were described as 'coffee houses' on board, though it appears that they also served other, more potent, beverages. There was both first and second-class accommodation and both boats ran to strict timetables. Like the fly boats, they had precedence over other craft and, unlike the fly boats, had a means of enforcing it. Each craft carried a curved, shining blade in the bows, which could be used to cut the tow rope of any craft that refused to give way. At first, they were simply referred to as passenger boats, but the name was soon changed to one already used for coastal boats – they became packets.

Travel by boat seems to have been quite popular, and we do have a long account of a trip on a Bridgewater packet taken by Sir George Head in 1835. Not surprisingly, he found a marked difference between travelling through the countryside and passing through the urban areas.

> This mode of travelling, to an easy-going individual, provided it be not repeated too often, is far from disagreeable; – there he sits without troubling himself with the world's concerns, basking in the sunshine, and gliding through a continuous panorama of cows, cottages, and green fields, the latter gaily sparkling in the season with buttercups and daisies … It is true, there is one drawback to the comfort of the traveller, – namely that within a dozen miles of Manchester the water of the canal is as black as Styx, and absolutely pestiferous, from the gas and refuse of the many factories with which it is impregnated.

There was a choice between staying in the cabin or sitting on one of the benches on the flat roof. Sir George was out in the open air, enjoying the view when the tranquillity of the scene was suddenly disturbed by a woman who fell head first off the top onto the lower deck. Sir George felt she must have been killed but went to her assistance. 'As it was, she was not hurt, and, as I picked her up, she sent forth a sigh, which smelt so strongly of rum that I was happy to consign her collapsed form into other hands.' Sir George fared no better when he took to the first class cabin. Although there is a good deal of snobbery in his description of the scene, one does have some sympathy. We have probably all at some time had to share a compartment with a doting family and howling child – even if not accompanied by a servant.

> The man and wife smirked and smiled on each other, and both gloated with eyes of affection on the dear baby. The lady, anxious to show to the rest of the passengers that she kept a maid-servant, ever and anon was calling for her from one part of the vessel to another to give her some trifling order. The little maid, nevertheless, seemed truly happy, and the more the child cried, the more she jiggled it, and the more her active eyes travelled round and round, looking first on one person then on another, while she sparkled with delight as she inhaled the pure air.

By this time, there were also packet boats operating on the Mersey and Irwell Navigation. Sir George also used this service and found the company even less agreeable. Not only were some of his fellow passengers somewhat rough-seeming to the aristocratic traveller, but they actually had the temerity to wander into the first-class cabin, drinking beer and smoking. He did, however, give a description of how the boats were worked. One might have thought that, given their high prestige and need to keep to a timetable, the job of hauling the boats along would be entrusted to fine horses, controlled by experienced men. The reality was very different.

> The boat was towed at the rate of about five miles an hour by a couple of clumsy cart-horses, driven beyond their natural pace, and working under all possible disadvantages, for half the strength of one horse was continually exerted to prevent itself from being dragged into the canal by the other … In the present case, the two small boys who rode each on one of these unfortunate horses, exhibited an utter insensibility to that lively state of muscle which is the result of a well-tutored mouth. They whipped and kicked as if sitting under a tree;

while the horses tugged and reeled, exhibiting a perfect specimen of ill-applied force, one literally pulling one way, and the other another. In the meantime, the riders, in worsted stockings, with thick country-made shoes, were healthy and active, jumping on and off, according to their fancy, without stopping the boat or creating any delay.

These two lads were, Head thought, no more than twelve years old, but they did the whole 32-mile journey in this way, every single day, returning the next, regardless of weather or season.

One of the new developments that marked life in the eighteenth century was the picturesque movement. This meant, literally, an enjoyment of scenery that would look good in a picture – not just any picture but in the very romanticised landscapes of artists such as Claude. Rugged mountains were now considered far more attractive than fields of corn. Where earlier writers had travelled through the Lake District and simply dismissed it as a barren area, it was now lauded for its splendour. The area had become a tourist attraction and the proprietors of the Lancaster Canal decided to cash in on the trade by running packet boats to Kendal from the industrial towns of Lancashire. Other companies followed suit. It was not just mountains that had become popular, the seaside had also become an attraction.

A crowded horse-drawn passenger boat on the Regent's Canal. Everyone seems to be enjoying the trip, apart from the stout gentleman in the middle.

Wigan Pier is usually thought of as a bit of a joke. There was a staithe here for loading coal, but there was also a landing stage for the packet to Liverpool, and passengers could also leave the boat at Halsall, where they would be picked up by coach and taken to Southport. Surprisingly, perhaps, the canals themselves proved an attraction in the early days of construction. The idea of going underground by boat was a novelty and the Reverend Stebbing Shaw took a trip on one of the pleasure boats that took visitors through the Harecastle tunnel, on the Trent & Mersey, shortly after it was opened in 1777. 'The procession was solemn; some enlivened this scene with a band of musick, but we had none; as we entered far, the light of candles was necessary, and about half-way, the view back upon the mouth was like the glimmering of a star, very beautiful.' The original tunnel – now replaced – was low and narrow, roughly nine foot wide. One can only imagine what a band would have sounded like in such a confined space.

The coming of the railways brought competition. Sir George Head, who had taken a whole day to travel thirty-two miles by canal, took a trip on the Leeds and Selby Railway soon after it opened in 1834. He recorded with commendable accuracy that he travelled 19 miles 7 furlongs in 1 hour 4 minutes. If the packet boats hoped to compete, they were going to have to improve on the performance of a bulky vessel dragged along by carthorses. As with many advances in the development of craft for the waterways, the answer was found by a Scottish engineer, in this case William Houston. He identified a problem with wide vessels in narrow canals. To make progress, water has to be pushed back down the sides of the vessel and if there is not much space for that to happen then a bow wave develops in front of it. Using faster horses does not help – as the speed increases, so too does the size of the bow wave, slowing everything down again. Houston began experimenting with narrower vessels, including one that was 60ft long but a mere 4ft 6in wide. That proved too extreme, but he found that by improving the lines of the hull, he could produce a vessel that could be pulled at up to twelve miles an hour. This was to be the basis for what became known as Scotch boats.

One of these vessels was sent down for use on the Lancaster Canal in 1839. She was 70ft long and 6ft beam and could carry

90 passengers. There was a first class cabin at the front, second class at the rear and a bar in between. One of these vessels, *Waterwitch II*, continued as an inspection boat on the canal, and survived right through to the 1930s, long enough to be photographed (below).

There was, of course, an alternative to the fast, horse-drawn boat – the steamer. This was obviously useful on wide rivers. In 1826, the first steamer to be built in Bristol, the *Wye*, was put through its paces on the run from that city, down the Avon, across the Severn and up the Wye to Chepstow. A voyage that had taken trows two days, was made in just 1 hour 55 minutes. She would later be joined by passenger steamers and in 1843, an iron-hulled steamer was offering passengers a fast and comfortable journey at three shillings single, five shillings return for a place in the cabin, or half that for anyone prepared to stay on deck. There was, however, still a case to make for extending steamer services to the canal system.

The Forth & Clyde, which had seen the trials of the *Charlotte Dundas*, was to be the scene for a new experiment in 1830.

The former Lancaster Canal packet boat *Waterwitch II* was built in 1839 and was kept on as an inspection boat but was finally broken up shortly after this photograph was taken in 1952.

The engineer Thomas Grahame designed a stern-wheel steamer, the *Cyclops*. However, according to William Fairbairn, the pioneer of steam navigation who saw the trials, it was not a brilliant design. This is his description of the trial from his *Remarks on Canal Navigation*, 1831.

> The boat, except as regards shape, is replete with errors. She is too heavy, viz. she bears about with her a quantity of iron, sufficient to build nearly two boats of the same size, and of equal strength. Her engine, which ought to have been high pressure, is low pressure, and, although a sweet-going machine, is much too heavy. Her paddle, which from its position must necessarily labour under the disadvantage of a deficient supply of water, is so placed so as to enjoy this disadvantage to its greatest possible extent.

In spite of this damning criticism, Fairbairn concluded that stern-wheeled steamers were ideal for use on canals. In fact, they were to have limited success. Steamers only really became practical as canal boats with the invention of the screw propeller, as described earlier. Several steamers were built for Scottish canals and they received a great publicity boost when Queen Victoria took a trip on the Caledonian Canal in September 1873 on board the *Gondolier*. She was not actually greatly impressed for although she praised the engineering, she found the journey itself 'very tedious'. She was, however, entertained by the scene when passing a lock. 'It was amusing to see the people, including the crew of the steamer, who went on shore to expedite the operation, run round and round to move the windlasses'. The 'windlasses' she referred to were capstans, used to open and close the heavy lock gates. It was enough, however, to help promote the idea that steaming through the Highlands was a good idea, and all the Scottish canals were soon running regular passenger services. It was not only the steamers' owners who hoped to make money from this new tourist trade. The Crinan Canal authorities, however, took a dim view of young entrepreneurs.

> Children and Others are hereby Prohibited from Running along the Canal Banks after the Passenger Steamer; and

The steamer *Gondolier* on the Caledonian Canal. This was the vessel on which Queen Victoria travelled when she visited the canal in 1873.

The pleasure steamer *Gypsy Queen* at Shirva on the Forth & Clyde Canal in the early years of the twentieth century.

Passengers are requested not to encourage them by throwing Money on to the bank.

Children are further warned not to throw Flowers onto the Boat.

As well as the regular passenger boats, there were market boats. This was a well-established part of river traffic. They ran to no fixed schedule, and were primarily used for cargo, but usually had a space covered with an awning for passengers. Many commercial carriers also had a lucrative side line in the passenger trade. Pickford's, for example, had a contract for carrying military personnel when necessary. A rather less savoury business was moving the poor and destitute. Under the Poor Law, families could only get parish relief in their original home parishes. If they had left, hoping for a better life, and failed, then they were shipped back. Some, more fortunate, were shipped to manufacturing areas if there was work on offer.

Salter Brothers have been running passenger boats on the Thames since the middle of the nineteenth century and still do so today.

120 • CRAFT OF THE INLAND WATERWAYS

The Thames Clipper service was begun in 1999 using fast catamarans, offering both a regular commuter service on the river in central London and pleasure cruises.

Eventually, the demand for passenger boats dwindled with the spread of the rail network. On the Thames, at least, they still have a role to play. Salter's Steamers have been running passenger boats on the Thames since 1868. They still run a fleet, although steam has now given way to diesel on the latest boats. These are essentially excursion vessels, not intended to take anyone from A to B, but simply allowing people to enjoy the river scenery. In the London area, however, boats still provide a valuable regular passenger service. The latest river craft are catamarans. The Thames Clippers went into service in 1999, but in 2020 were renamed Uber Boats. They offer a fast commuter service between Westminster and Woolwich Arsenal Pier.

Chapter Eight

Ferry Boats

Ferries have been an essential part of river life since ancient times. In Greek mythology, Charon is the ferryman who carries the souls of the dead across the River Styx. Rivers have always been valuable as transport routes, but for land transport they represent barriers to cross. Bridges are expensive to build, so towns often developed around places where they could be crossed on fords. Surprisingly, it was once possible to cross the Clyde in the centre of Glasgow on horseback. But fords were few and in many cases the only option was to make a long detour or take a ferry. Until the middle of the nineteenth century, anyone wishing to travel from Bristol to Chepstow in South Wales by road would have to go north of Gloucester to find the first bridge and then go back down the opposite side of the river, a round trip of over seventy miles. It is hardly surprising that a ferry would have seemed a better option. In other cases, it was the sheer volume of traffic and lack of bridges that created the problem. As we have already seen, London was plagued by the lack of bridges over the Thames for centuries and ferries had their own part to play, especially for anyone wanting to take a vehicle over the river. But ferries were not limited to the city – they existed up the Thames above Oxford. The upper Thames was traditionally fordable, but there was a working ferry at Bablock Hythe, half a dozen miles north of Oxford. Walter Armstrong came here c.1870 as part of a journey along the river, which he described in two large volumes, *The Thames from the Rise to the Nore*. Fortunately, Armstrong was also an accomplished artist and he has left us a whole series of excellent engravings of what he saw along the way, including the Bablock Hythe ferry.

Ferries are unlike other river craft in that they have a very limited function, simply to take passengers from a fixed point on the river bank to another fixed point opposite. They could either run a regular service, or wait until they were summoned,

Bridges were sparse on many rivers until the nineteenth century and often fords were essential crossing points as here on the Thames above Oxford.

often by nothing more complicated than ringing a bell. One of the problems faced by ferrymen is the need to make a straight crossing from one bank to the other, while the current is trying relentlessly to push them downstream. Skilled oarsmen could cope, but here a different technique has been used. The flat-bottomed boat has two vertical frames, holding rollers. A rope secured on both banks, and tightened when necessary by a windlass, is threaded through the frame, running past the rollers. This ensures that the boat cannot drift downstream. The ferryman then hauls the boat across, using the overhead line – in the illustration one can see that this is definitely hard work.

Running a ferry at a popular crossing point could be a lucrative business, particularly if the owner had a guaranteed monopoly. But having a legal right did not necessarily mean that everyone respected it and didn't try to muscle in on the trade, as the story of a ferry across the Thames at Twickenham illustrates. The largest landowners in the area were the Earls of Dysart, who owned the magnificent Ham House, built in the seventeenth century. The earliest records of a ferry are in documents of 1659, when the Dysart family held the exclusive rights – though

Where there were no fording points on the upper Thames, ferries were introduced. This example was in use at Bablock Hythe at the end of the nineteenth century.

there is, of course, no way of knowing if there was an earlier, undocumented ferry. In 1692, the service was leased out to William Blower at a rent of forty shillings per annum. He died shortly afterwards in 1692, but his wife carried on the business until her death ten years later. The lease was then taken on by Thomas Love and once again he died early and his wife again carried on until 1744. They must have been formidable ladies.

The eighteenth century saw the first attempt to develop a rival business. Two men, Treherne and Langley, set up a ferry very close to the Dysart. They were immediately taken to court, lost the case and were fined sixpence. It was not exactly a massive fine, but it was obvious that if they tried again, they would be back in the dock and receive considerably harsher treatment. However, another of the enterprising ladies who are such a feature of this story, Margaret Langley, widow of the earlier ferryman, went into partnership with Samuel Kain. This time, however, they came to a legal arrangement with the Dysart family, providing a £100 bond, and was given permission to operate a 'publick ferry', provided she did not interfere with the original Dysart ferry.

Throughout the eighteenth century, Twickenham and the surrounding area had been largely rural but in the nineteenth century it was being developed and houses were being built for the newly prosperous middle classes. There was an early environmentalist movement that wished to save such valuable green spaces within easy reach of London from suburban sprawl. The National Trust was formed in 1895 to preserve ancient buildings and to protect the countryside. They campaigned on behalf of the green area by the Thames and as a result the Richmond, Petersham and Ham Open Spaces Act was passed covering a large area of the Surrey bank. At the same time, the London County Council acquired the Marble Hill estate on the opposite bank, which was opened to visitors. This brought new trade to the Twickenham ferry, though it was not very close to Marble Hill. The ferry man at the time was a former Dysart employee called Champion who had been appointed in 1891 when he and his wife moved into the new Ferry Cottage. Mrs Champion proved as enterprising as her predecessors, renting a property by the ferry landing on the Surrey side at £1 a week, where she opened a café. Business prospered and they soon had three boats, each able to take twenty passengers, and hired a 'strong boy' to help out. They were simple rowing boats intended solely for the tourist trade.

The Twickenham ferry – as well as providing a ferry service, the ferryman's wife had a tearoom in the house on the left.

It was far too successful to go unnoticed, and others once again decided to join in and gain a share of the business. A licensed waterman, Walter Hammerton, was given permission by the Council to moor a floating boathouse near Marble Hill, from which he rented out skiffs and punts. There were riverside steps directly opposite, so anyone who arrived on the opposite bank had only to wave and someone would row over and fetch them. That was all perfectly legal, but soon they were rowing across to fetch people who did not want to hire a boat, but simply get across the river. The boat hire business was becoming a ferry service. He might have got away with it, if he hadn't painted a sign saying 'Ferry' on the boathouse. Once again, it all ended up in court. On the first hearing, Hammerton won, but the Dysart estate took the matter to the Court of Appeal and this time they won. Now it was Hammerton's turn, and he went to the House of Lords, and there was another reversal. Now both ferries could run legally, and both made a profit. The court cases' main beneficiaries were the lawyers – it could all have been settled so easily, if only the two parties had not been so stubborn.

The Hammerton ferry still operates, and another nearby ferry also survived for a long time to serve Eel Pie Island, which apart from a small number of residents also had a hotel. It was to become a popular music venue in the 1920s, when it had a resident dance band and enjoyed an even greater following post-war, first for traditional jazz in the 1950s then with rock bands including The Who and the Rolling Stones. The hotel, however, became more and more dilapidated over the years and was burned down in 1961. The island can now be reached by a footbridge.

These were people ferries, but in the heart of London, there was also a demand for ferries that could take horses and vehicles. One of the most important crossings was the Lambeth horse ferry – which, although it no longer exists, is still remembered by the name Horseferry Road that now leads directly to Lambeth Bridge. It is long established, with references going right back to the thirteenth century, when there was a dispute about charges between Lambeth Palace and St. Peter's Abbey in Westminster. Because of the strong currents and tides, it was necessary to have some means of keeping it on course, so a more sophisticated version of the Bablock Hythe ferry was in use. Instead of being pulled across by a cable, it is moved by using a chain and sprocket. The chain is kept quite slack, so that when not in use it lies low in the water so as not to impede other craft. It is fed into the boat, where it engages with the sprocket, which can be turned by hand by an operative on the deck using a winch. It can be seen at work in the illustration below. The boat itself is flat bottomed, with gently sloping bow and stern for easy access.

In spite of being attached to a chain, accidents did happen. In 1633, the ferry sank when it was taking Archbishop Laud, his servants and horses across. Everyone escaped, but the Archbishop's good fortune ended in 1645 when he was beheaded for treason – a crime for which there was scant evidence. Oliver Cromwell was another victim of a capsize in 1656 and although he survived, the horses that were being carried went down with the carriage and drowned.

The ferry brought a very handsome income for Lambeth Palace, who leased out the ferry, and to those who operated it. Fares in the seventeenth century varied from twopence for

The Lambeth Ferry by Jan Groffier. This was one of the few in London able to take carriages over the river. It was operated by a chain, and the ferryman is winding the boat along it.

a man and horse, to two shillings and sixpence for a coach and six. Not surprisingly they were among opponents of any attempts to build a bridge. The first attempt to get authorisation to build a bridge was made in 1664, but it was not until 1756 that Parliament approved construction and then the Archbishop was awarded £3,780 as compensation for loss of revenue. In terms of purchase power that is nearly £800,000 at today's values, and the ferry didn't even close down. The Lambeth bridge only carried wheeled vehicles, so riders and pedestrians still relied on the ferry. It finally closed when the first suspension bridge was completed in 1862.

As more and more bridges were built in the London area, so the ferries gradually disappeared, but there were no bridges south of the Pool of London until the opening of the M25 orbital motorway. One ferry that started serving a purely rural area at Woolwich grew in importance with the development of the Royal Arsenal. By 1810, the army had their own ferry service and for a time a private company operated as well. That collapsed in 1846 and at this point a new owner came on the scene, offering a very different kind of service – the Eastern Counties and Thames Junction Railway. They built a branch line down to the riverside in 1846 and they naturally turned to steam power for the ferries. Three paddle steamers were built for the crossing. By 1880, the service appeared to have become very run down, and attitudes towards river crossings had changed. The first generation of bridges had all charged tolls, but they had been abolished by law. The citizens of Woolwich and London's East End saw no reason why, when other Londoners could cross the river for nothing, they still had to pay. The Metropolitan Board of Works agreed, and a free ferry service was begun with new steam ferries. Steamers no longer do the work but the ferry still runs and is still free.

Steam came into use in other parts of the country as well. Ferries on the Humber date back to the early nineteenth century, but as in London, the coming of the railways saw the introduction of steam. In the 1930s, paddle steamers were built, two of which, both constructed in 1934, *Tattershall Castle* and *Winsfield Castle*, have both been preserved as floating restaurants and bars.

The Woolwich steam ferry. Passengers travelled for free on this public ferry and they still do on its successors.

Paddle steamers were used for the ferry service across the Humber from Hull. One of these vessels has been preserved. *Tattershall Castle* is now a pub by the Thames embankment.

Few ferry crossings were more dangerous and difficult than that across the Severn near the mouth of the Avon. The river has the second highest tidal rise and fall in the world – the highest is the Bay of Fundy off the coast of Nova Scotia. There were two fatal accidents to ferries, first in 1839 and again in 1844 – by a macabre coincidence the ferry captain who died in the second tragedy was the son of the captain who perished in the first. The age of the motor car brought a new car ferry service operated by the Old Passage Ferry Service in 1931. This used motors instead of steam power and had an unusual feature. Instead of loading cars at one end, they boarded via a ramp at the side and were stopped on a turntable, swung through 90° and lined up. Nineteen cars could be packed in. The service continued with three vessels, the *Severn King*, *Severn Queen* and *Severn Princess*. Even at this late date, the river claimed victims. In October 1960, two barges collided in thick fog, and drifted onto the piers of the old Severn Valley Railway bridge, near Sharpness. One of the barges had a cargo of petrol and exploded, killing five crew members and causing such severe damage to the bridge that it had to be demolished. The *Severn King* was sent to help with the demolition work, but also became a victim, when she settled on one of the ruined piers and broke her back. The ferry service continued until September 1966, when the first of the Severn bridges was opened.

One ancient ferry has survived through the centuries. The King Harry Ferry crosses the River Fal three miles south of Truro. The local story is that it was given a charter by Henry VIII when he was on his way to St. Mawes with his new wife, Anne Boleyn. But the name King Harry goes further back. There was once a church here – the chapel of St. Mary and King Harry – and in this case the king was the ill-fated Henry VI who was killed in 1471. Like all ancient ferries it has gone through several transformations. In 1888, a steam-powered chain ferry was established and although it is still a chain ferry the present version is powered by a diesel-electric system. It is popular, not because it saves a huge distance for travellers, but simply because it offers a chance to cross the river at a particularly beautiful spot.

Few ferries survive these days, but in their time they were an essential part of the life of Britain's waterways and vital

The Severn car ferry was a vital service in the days when there was no road bridge lower down the river than Gloucester. But the photo also shows the new bridge under construction that would end the ferry service.

The King Harry chain ferry across the River Dart is supposed to have been named after Henry VIII who is said to have ordered its construction.

FERRY BOATS • 131

links in the whole transport system. They show as much diversity as any other river craft and sometimes looking at old pictures one could wish that more had survived, just for the sheer pleasure of enjoying a simple little trip across the water in a picturesque setting.

A scene that was once typical of rivers all over Britain – the Norwich ferry from an old postcard.

Chapter Nine

Boating for Pleasure

There is little real evidence of pleasure boating before the eighteenth century. It was then that the cult of the picturesque came into being, and one of its main proponents was the Rev William Gilpin. His *Observations on the River Wye* of 1770 was an influential best seller, and the following passage, describing a boating trip on the river, sums up his whole attitude:

> After sailing four miles from Ross, we came to Goodrich Castle; where a very grand view presented itself; and we rested on our oars to examine it. A reach of the river forming a noble bay, is spread before the eye. The bank on the right is steep, and covered with wood; beyond which a bold promontory shoots out, crowned with a castle, rising among the trees. This view, which is one of the grandest on the river, I should not scruple to call *correctly picturesque.*

The italics are Gilpin's, emphasising that the picturesque had strict rules. One might have thought he would have rhapsodised over the Wye's most famous building, Tintern Abbey, but that did not meet his standards of just how ruined a romantic ruin should be. 'A number of gable-ends hurt the eye with their regularity; and disgust by the vulgarity of their shape. A mallett, judiciously used (but who durst use it?) might be of service in fracturing some of them'. So much for Tintern. Happily, no one did try and knock bits off. But he did a great deal to encourage people to think of rivers as attractive places on which to spend one's time. The nineteenth century was to see an upsurge in interest in exploring Britain's rivers, and boating on them, and no river was more popular than the Thames.

By the end of the nineteenth century, there were all kinds of pleasure boats on the river, many such as punts and rowing boats simply used for short day trips, but there was also a demand for boats that could be used for holidays that could ideally travel the length of the non-tidal river. The classic account of this type of holiday is Jerome K. Jerome's *Three Men in a Boat*, first published in 1889. The three hired a camping skiff. This was a large, clinker-built boat, with two pairs of oars and steering via a pair of lines attached to the rudder. The seats for rowers were fixed, not sliding. If the wind was favourable, there was a short mast and a single sail available on some skiffs. At night, it could be turned into sleeping accommodation by setting up a set of five iron hoops to hold a canvas cover over the boat. In Jerome's account, what should have been straightforward, taking, they estimated, ten minutes, became something of a comic nightmare. This is part of the account.

> We took up the hoops, and began to drop them into the sockets placed for them. You would not imagine this to be dangerous work; but, looking back now, the wonder to me is that any of us are alive to tell the tale. They were not hoops, they were demons. First they would not fit into their sockets at all, and we had to jump on them, and kick them, and hammer at them with the boat-hook; and when they were in, it turned out they were the wrong hoops for those particular sockets, and they all had to come out again.
>
> But they would not come out, until two of us had gone and struggled with them for five minutes, when they would jump up suddenly, and try and throw us into the water and drown us. They had hinges in the middle, and, when we were not looking, they nipped us with these hinges in delicate parts of the body; and, while we were wrestling with one side of the hoop, and endeavouring to persuade it to do his duty, the other side would come behind us in a cowardly manner, and hit us over the head.

It took them half an hour to finally get the cover in place. Obviously, Jerome is exaggerating their struggles for comic effect, but reading the account made me glad that when my wife and I made a similar journey, by canoe rather than skiff, we slept in a tent. However, camping skiffs are still available for hire on the river. They are comfortable, easy to handle and the covers

A busy scene at Boulter's lock on the Thames in the early twentieth century. In among the boats are punts, skiffs and, at the far end of the lock, a steam launch.

can be quickly put in place so none of the problems encountered by the three hapless men are likely to be repeated.

The rowing boats and punts were not the only craft enjoying the river when Jerome and his friends made their voyage. Steam launches had also become popular. Jerome was at first firm in his hatred of the craft. 'There is a blatant bumptiousness about a steam-launch that has the knack of arousing every evil instinct in my nature.' That was the case, at least, until they met friends with a launch who offered them a tow – then he complained bitterly about how all those wretched rowing boats kept getting in their way.

The oldest surviving steam launch was built around 1850 for use on Ullswater in the Lake District. *Dolly* is a modest craft, 32ft long by 5½ft beam. Her simple steam engine has a 7in diameter cylinder, built by a local blacksmith, and she is driven through the water by a screw propeller at a modest 5 knots. She pottered around the lake quite happily for years, but on 21 February 1895, a young woman was looking out of her window watching the little boat puff along. She turned away for a moment, and when she looked again it was nowhere to be seen. *Dolly* had sunk. That would have been the end of her story, but she was found sixty years later by divers for a local sub-aqua club. In 1962, she was raised from the bed of the lake and taken to Windermere. Gradually the pine and oak of her hull were dried out and the time came to test her engine. Being in fresh water, the iron had not corroded as it would have done in salt water and when she was tried out, the engine worked just as it had when she was built. She is now one of the prize exhibits among a fine collection of launches at the Windermere Jetty Museum.

Alfred Yarrow as a young man had been on a steam launch trip and enjoyed the experience. He had shown himself to have an inventive mind from childhood. His first invention was a very practical device that saved the boy from boredom. He had an aunt who was a keen knitter, and the boy was constantly being called on to hold skeins of wool for her, so he made an automatic wool winder. He went on to other devices, including an automatic candle snuffer and set up a primitive electric

The simple single-cylinder steam engine of *Dolly*. Built c.1850 she is the oldest surviving steam launch.

telegraph to a friend's house. As a young man, he worked on a steam plough with a friend called Hilditch, and developed a steam carriage, but the notorious Red Flag Act, limiting the use of steam on the roads, made that less successful than he had hoped. In 1863, at the age of 23, he acquired a boatyard on the Thames at the Isle of Dogs, with another friend, Hedley, and set up in business as a ship repairer. In the first year they lost £100 and the next year business increased, but the losses rose to £2,000. It was then that he decided to try a new line of business and put out an advertisement that read, 'Steam Launches – anyone wanting a steam launch would be well served if they came to Yarrow and Hedley, Isle of Dogs.'

They got one customer at first who paid £140 for a launch that cost £200 to build. But Yarrow managed to buy it back shortly afterwards for £100 and sold it for £200 – so eventually ended up with £40 profit. He then set about using the launch in an advertising campaign. He took a photograph of it and had a large number of copies made, which he stuck up in waterside pubs and inns up and down the river. Between 1868 and 1875, when the partnership was dissolved, they sold 150 launches. Yarrow would go on to build far bigger vessels and would eventually design and build torpedo boats for the Royal Navy. Steam launches were obviously successful, and another famous name in ship building, John Thornycroft, also began his career building launches on the Thames at Chiswick.

The appeal of the steam launch is obvious. The museum on Windermere where *Dolly* is housed is home to a fine array of these elegant vessels. My wife and I were invited by George Pattinson, whose collection was the basis for the museum, to join him on *Kittiwake*, a vessel that epitomises the elegance of these craft. She is a sleek vessel, 40ft long overall and 7ft beam, constructed of varnished wood. There is a comfortable cabin near the stern. When looking at earlier steam engines, one of the most complex was the VIC32 – a compound engine that uses steam pressure at a maximum pressure of 120psi. *Kittiwake* goes one better, with a triple expansion engine, in which the steam at 200psi first enters a 5in diameter cylinder, then exhausts to a 6¼in before finally passing to an 8in cylinder. This makes for very economical running, and the fuel used is wood. Mr. Pattinson told me how, when he was a boy, they would set out for a picnic with just enough wood on board for the outward journey, and the children would have to find enough branches and twigs lying around for the return. He also demonstrated one of the special features of such launches, the Windermere kettle. This is basically an urn full of water, with metal coils inside. When a valve is opened, steam from the boiler rushes through the coils, boiling the water in seconds. Perhaps, the feature that does more than any other to make a trip by steam launch popular is that the engine is so quiet and progress so peaceful. That was about to change in the twentieth century with the arrival of the motor launch. But before looking at how that came to dominate popular river boat travel, we'll step back in time to see how traveling for pleasure on the canals first came about.

Right at the beginning of the canal age, Josiah Wedgwood and Thomas Bentley brought out a pamphlet making the case for the construction of the Trent & Mersey Canal in 1765, in which they rhapsodised on the delights of having a canal at the bottom of the garden – 'a lawn terminated by water' – and even suggested that the lucky house owner might like to have a gondola there

George Pattinson and his family on the steam launch *Kittiwake*. His collection of launches was the foundation for the present steamboat museum on Lake Windermere.

for his own use. No one, it seems, took the idea seriously. The Trent & Mersey was then far too busy with commercial traffic to make pleasure boating an appealing option. One of the earliest references to pleasure boats on the canal came in a magazine article of 1885, describing a trip on the Regent's Canal. 'Small pleasure-boats are allowed to ply on parts of the canal, and have given life to the scene. A large barge is anchored in front of a green field, and its owner informs us by a sign that he has "boats to let for school and picnic parties."' There is no indication of what the boats were like, but an earlier illustration does show a simple rowing boat on the canal, with the passengers seeming to overflow it (below).

These were boats offering short trips and it seems they were popular on some of the more picturesque canals, and few are more picturesque than the stretch along the feeder canal at Llangollen that ends at the great Pont Cysyllte aqueduct and, as the photo (page xxx) shows it was just as popular a century ago as it is today. But there are few mentions of boating holidays on the canal system until the late nineteenth century. *Two Girls on a Barge* by V. Cecil Coates, published in 1891, is just what the title suggests, except that the barge was a narrow boat. They arranged to hire a boat and crew, the latter referred to only as Mr and Mrs Bargee. They then had the hold divided up into cabins, which the young

Rowing boats were available for hire on the Regent's Canal in the early nineteenth century. The gentleman and his ladies are being rowed by a hired boatman.

A mixed collection of pleasure boats on the canal at Llangollen early in the last century. Horse drawn trip boats are still available for tourists to enjoy a voyage down to the Pontcysyllte aqueduct.

ladies then proceeded to decorate in the 'arts and crafts' style that had become popular at the end of the Victorian era.

There was still much to be done before we could begin to take Barge life in earnest. There were the curtains to hang. Liberty curtains that had taken a whole day to choose, and 'dhurries' to be draped over the fresh painted pine of the little cabins, and Liberty again in innumerable hangings to be arranged all round the bulwarks gracefully.

What Mr and Mrs Bargee thought of all this as they sat in the little cabin that was their only home is not recorded. They do, however, seem to have taken some pleasure in telling the women terrifying stories, which they recalled when going through Blisworth tunnel on the Grand Union.

We went in a black, domed passage, and it was deathly still; in all that string of barges no one spoke or moved. The gloom of

One of the earliest accounts of pleasure boating on Britain's canals is *Two Girls on a Barge* written in 1891. Here they are with a friend by the Watford staircase on the Leicester Arm of the Grand Union.

the place encircled us. An indefinable Presence moved with us in the blackness. The nearness of the damp stones impressed itself upon the eyes and played fantastic tricks with the imagination. Every sense became distorted, unnaturally acute; the silence was appalling. The story the Bargee had told us of the great White Spirit, boding evil to the boatmen whenever it appears, came back to me with a meaning and terror that yesterday had seemed impossible. 'Ye just slips yer foot, or over balances, and the black water swallows ye.'

This would seem to be the first account of adapting a working narrow boat for use as a floating holiday home. In another early account, the classic *The Flower of Gloster* of 1911, the author, E. Temple Thurston, simply buys the narrow boat of the title and hires a boatman for a canal holiday, but does not appear to have bothered to have had special cabins made. He was happy just to enjoy the peace of the countryside and the slow progress.

There is a photo of a party of gentlemen taking a steam launch along the Leeds & Liverpool in the 1890s (page 141) but I have

A rare **photograph** of a party of gentlemen enjoying a journey by steam launch on the Leeds & Liverpool Canal in the 1890s.

been unable to find an account of their journey. Steam launches were, however, used by canal managers for inspection purposes and Henry de Salis, chairman of Fellows, Morton and Clayton, was one of the early users. He is probably better known as the compiler of the Bradshaw *Handbook of Inland Navigation* that provided information on every navigable waterway in Britain. As he announced on the title page, he personally surveyed every single mile.

By 1915, a very different type of craft was appearing – the motor cruiser. P. Bonthron's book *My Holidays on Inland Waterways* describes how he and friends travelled in a motor boat powered by a 6hp engine, though once again they had a paid crew to do most of the work:

> Our motoneer [a cognomen suggested at the time for a motor-boat engineer] and the handy man – both being in great demand for the many duties to be performed, the latter being

much sought after at times as the provider of all good things in the way of light refreshments, etc.

They also represented a phenomenon that most waterways enthusiasts will have come across at some time or other, the 'lock basher' who seems less interested in the scenery than in how far he can go and how many locks pass in a day. Here he describes a memorable day's journey, starting on the Stratford Canal. They then moved on to the Warwick and Birmingham. They made a brief stop at Warwick, but soon hurried on again. 'In doing a trip such as this one feels inclined to land at many places en route to view the cathedrals and explore the town thoroughly but we found to adopt this would entail a serious loss of time.' The final leg of that day's journey brought them to the Warwick and Napton, where they passed Leamington without stopping.

We were really rather pleased with our run on that day, as altogether we had done 25 miles over part of the three canal companies' properties and the very large number of 66 locks –

Steam launches were also used by canal officials. This launch was used by Henry Randolph de Salis, chairman of Fellows, Morton & Clayton, who travelled over the whole inland waterways network of England and Wales to compile the canal Bradshaw guide.

an excellent piece of work we considered – constituting, as far as we know, a record for a single day's lock work.

There were more individual lengthy canal trips over the following years, notably that of L.T.C. Rolt in a converted narrow boat immediately before the start of the Second World War, described in his classic book *Narrow Boat*. Although boats for hire were already common on the rivers, it was only after the war that canal boat hire became a regular business. At first, many of the craft were simply river cruisers of an appropriate size, usually wooden cabin cruisers. A new form of boat also appeared in those years, with a fibreglass hull. By this time the canals had been nationalised and one of the biggest hirers was the newly formed British Waterways Board (BWB). For many coming to this type of holiday for the first time, it was the familiarity of the old-style craft that made them seem appropriate. When my wife and I took our own first

An early example of a working narrow boat being converted as a trip boat on the Basingstoke Canal, from an old postcard.

The frontispiece of P. Bonthron's *My Holidays on Inland Waterways* published in 1916, the earliest account of such journeys by motor boat.

canal holiday with BWB we hired a tiny cabin cruiser with an outboard motor called a Water Baby. It skittered around with every breeze – it sometimes felt that if someone sneezed on one bank it would blow us to the other. We did not repeat the experience on any other trip. Traditional narrow boats had been adapted for some time, by simply replacing the cargo area with cabins for long trips or with a single cabin for short excursions. It was obvious to some boat builders that a design that had been developed over a long period as ideal for canal travel would be the perfect model for pleasure boats. As a result, a new generation of hire boats was starting to appear, directly based on the working narrow boat, steel hulled and with the same tiller steering as found on a working motor narrow boat. Boats of this type are now ubiquitous and dominate the scene. It is fair to say that thanks to the growing popularity of canal holidays, there are now more narrow boats on the canal than at any time in the last 100 years.

An advert from the 1950s for canal holidays. At this time, hire companies were mostly using boats based on river cruisers rather than on narrow boats.

British Waterways ran their own hire boats in the 1950s including the two-berth, fibreglass Water Baby series, with outboard motors.

CHAPTER TEN

Building the Boats

The oldest form of construction for wooden barges was clinker building, in which the planks of the hull overlap. This was the method used, for example, for almost all wooden wherries, and the following description gives the main elements that would be employed for many other types of sailing barges. The one exception to this rule just happens to be the one featured earlier, *Albion*. She is carvel construction, with the planks abutting each other. There were small boatyards spread throughout the area, some such as Allen's of Coltishall and the Wrights of Beccles produced many wherries over the years. The last trading wherry to be built, *Ella*, was launched from Allen's yard in 1912. Other yards might only ever produce a couple of vessels.

The process began with selecting the timber, usually done by a master craftsman, who would visit the woodlands to select appropriate trees, usually oak. The great advantage of the oak is not just its toughness, but the way in which it grows, with branches coming out almost horizontally from the trunk. This means where curved pieces such as the knees that are used to connect beams to the hull – much as bracket supports shelf fastened to a wall – can be cut from the intersection and the grain will run true. Such pieces are known to shipwrights as compass timbers. Once the trees have been felled, they would be taken to the yard to be sawn in the saw pit. The trunk was rolled over the pit and secured in place. The lines where cuts were to be made were marked by string, and then cut using a double handled saw. One man stood above the pit, pulling the saw upwards, and the other, the pit man, was down below, pulling the saw down. The saw itself would have been about 7ft long, and the bottom handle was detachable, so that the saw could be pulled out when the blade had to be removed to move the log. The top sawyer had the harder task, but the bottom sawyer had to put up with being covered in sawdust throughout the working day. In hot weather, the men's armpits would become sore with sweat and they would rub them with fuller's earth.

In a clinker built vessel, the outer shell of the hull is completed first, the shape being determined by the use of moulds made of thin wood. These represented just half of the hull but had only to be turned round when it came time to construct the other

A scene that would have been common at any boatyard making wooden boats; two men sawing a log at a saw pit.

half. The hull of a wherry is curved, but planks are straight. To overcome this problem, the timber has to be heated to make it flexible. One rarely gets the opportunity to see this done but I was fortunate to be present during the building of a replica of the wooden carvel-built paddle steamer *Erik Nordwall*. The original was built in 1835 for use on the Göta Canal in Sweden. A plank would be placed in an open-ended chest, and steam from a boiler fed into it. This made the wood sufficiently pliable to be bent to the frame, after which it was firmly clamped to ensure it retained its shape as it cooled and was then nailed to the frame. Many of the wherry builders used a slightly different technique, using a bundle of burning reeds placed close to the hull to produce the same effect.

It was only when the strakes, the planks that made up the sides of the hull, were all in place that the frames were added. In carvel construction, everything is done the other way round. The frames are set in place, and the strakes attached afterwards. Shaping the timbers was mostly done with an adze, the archetypal shipwright's tool. This can best be described as being a bit like a long-handled axe with a curved blade, but one on which the blade has been turned through 90 degrees. When swung against timber, it automatically cuts a curve. The hull is made watertight by caulking. This involves pushing oakum, usually made from old ropes, into the space between the strakes using a special caulking tool with a triangular blade and a caulking hammer. The importance of caulking was brought home to me on my

A traditional Norfolk wherry under construction at Carrow in the early nineteenth century.

first trip on *Albion* when I shared the aft cabin with the skipper. The cabin roof had recently been replaced, but not yet caulked. It started to rain in the middle of the night, and a steady series of drips began to fall onto the bunk. It was impossible to find a way of lying down without some of them soaking us both, so there was nothing for it but to find a dry spot between the drips and spend the rest of the night chatting and making cups of tea. The masts on wherries were largely made of spruce.

Records survive of one typical small yard on the River Tamar at Calstock. James Goss was born in 1848 and began his working life in his father's shipyard at Barnstaple, but after a family quarrel left and went to sea as a ship's carpenter. He served for seven years, and then settled down to work at Edward

A group of Welsh shipbuilders. The man on the left is holding a caulking hammer and the one next to him an adze.

Brooming's yard on the Devon bank of the Tamar. When Brooming retired in 1878, Goss achieved his ambition of being his own master and took over the yard, to repair and build barges. It has been described as a set of tumbledown sheds on a mudbank. In fact, there was rather more to it than that. There were, in fact, two building sheds, two storage sheds – one for a variety of different gear and the other for rigging – and an office. There was a pitch house for heating coal tar, a saw pit and a blacksmith nearby. Goss never learned to read or write but relied on his own innate sense of right proportions. He drew the outlines for frames in pencil on a board he kept behind the kitchen door, and then took them to the mould loft. Accurate measurements were seldom if ever employed. Yet he produced a number of fine vessels, including several barges, most of which were employed in the unglamorous trade of carrying manure up and down the Tamar. His grandest vessel was the *Garlandstone*, a 150-ton ketch, built in 1909. She has survived and now has a permanent home at Morwellham open air museum. Like all craft coming from Goss's yard, she was launched fully rigged.

Although none of Goss's Tamar barges survive, one built by another yard does. *Shamrock* was constructed in Plymouth in 1899 by Frederick Hawke and is the last of its kind ever built. She is flat bottomed so that she could load even if no quay was available by simply being beached. She has a very rounded bow and square transom stern. Originally, she had two drop keels that could be lowered when necessary. She has two masts and is gaff rigged, though a bowsprit and jib were added later. In 1974 she was acquired by the National Trust and restored as a joint project with the National Maritime Museum. Her restoration involved using very traditional materials including hemp rope for the rigging. I filmed on board her in the 1980s and although admiring the authenticity found the ropes extremely uncomfortable to handle – I imagine one would get used to them in time. She is now berthed at Cotehele.

Larger yards would have their own sail lofts for making the sails. Some years ago there was a repair yard for Thames barges at Milton Creek, Sittingbourne, Kent that had a sail loft that was still in regular use, making sails in the traditional way. A sail is more complex than it might appear; it has camber. That is to say,

The Tamar barge *Shamrock* at Cotehele quay. She has been fully restored using traditional materials.

152 • CRAFT OF THE INLAND WATERWAYS

The sail making loft at the former barge maintenance yard on Milton Creek near Sittingbourne. The sail maker is stitching rope to the sail edge for extra strength.

if you had a square sail, say 10ft by 10ft, the actual area would not be 100sqft but slightly more. The difference is the camber, the curve. The sail maker has to work with tough material such as canvas and string thread and has a special leather palm for his hand. The sail would be made out of strips of canvas sewn together to make the appropriate shape and then rope would be stitched round the edges to strengthen it, and to provide the clews, used to attach the sail and control it. There were some

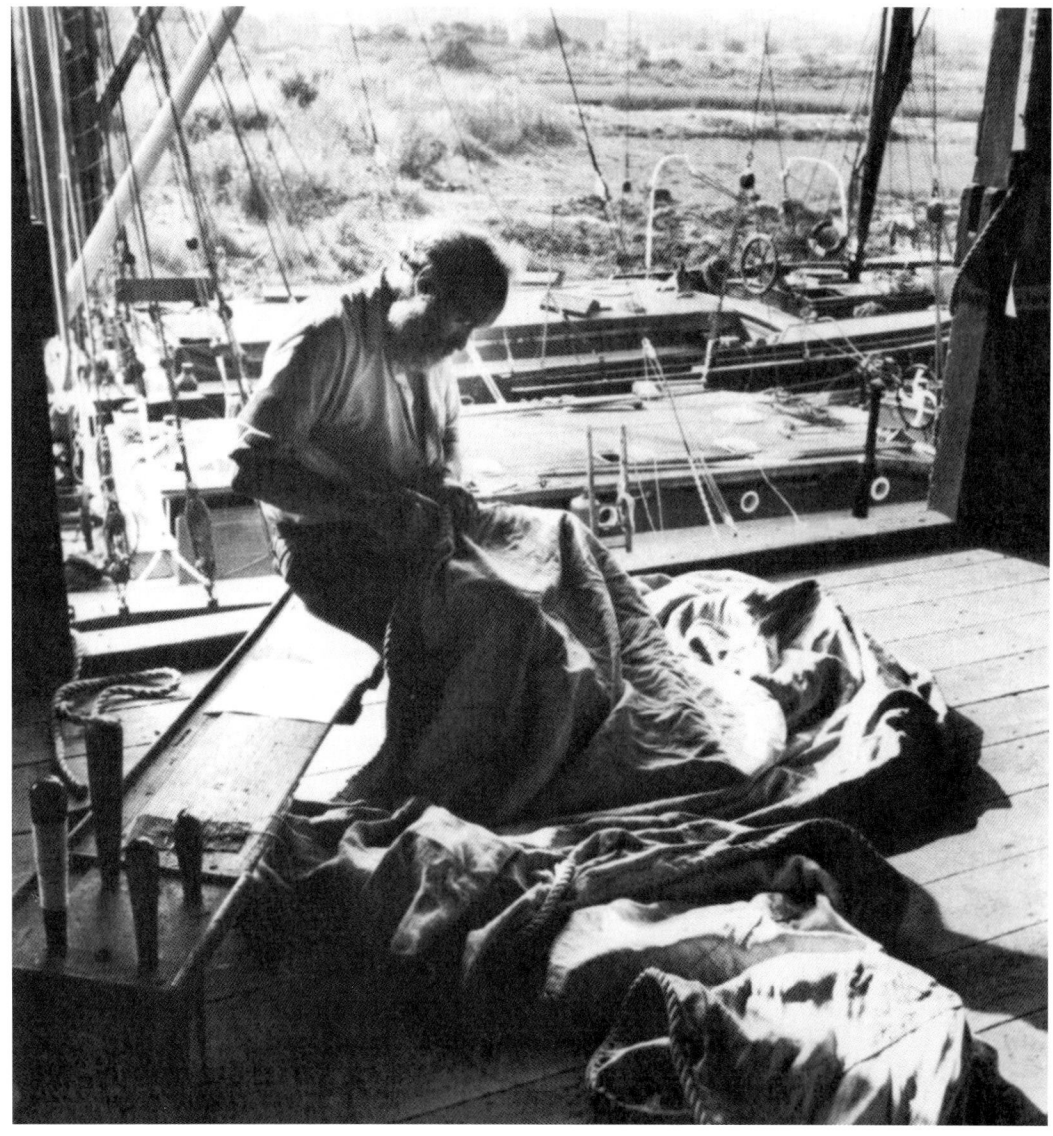

items that had to be bought in from specialists, notably rope for the rigging and the blocks that control it.

Traditionally, ropes were made from hemp. The first step is to clean and disentangle the fibres so that they lie straight and parallel, a process known as hackling. The hackler takes a bunch of hemp, a streak, lightly oils it and then pulls it through a hackle, a set of spikes mounted on a block. This may be repeated through several hackles, each with the spikes set closer together and mounted on a wooden block. When all the fibres are aligned, the hemp is ready for spinning in the rope walk. At one end of the rope walk is a wheel with four hooks. The spinner wraps the hackled hemp round his waist, and teases out some fibres, twisting them together with his finger and thumb, and attaches them to one of the hooks, repeating the process with the remaining hooks. He then proceeds to walk away down the rope walk, while a boy uses a crank to turn the wheel at a steady rate. The twisted fibres are laid over T-shaped wooden stands as he goes. In a quarter of a day, the spinner was expected to produce 160 fathoms (960ft). When the right amount had been spun, the yard was taken off the hooks and attached to a reeling machine, and the spinner would walk slowly back, winding the rope on. Inevitably, in time the process was mechanised, and a splendid example is still in use at the former Royal Navy dockyard at Chatham with machinery dating back to the early nineteenth century. The building itself is a quarter of a mile long, and instead of a man walking away from the turning wheel with its hooks, the wheel is mounted on a carriage, which moves away from the fixed frame holding the fibres.

Blocks are basically pulleys with a curved outer wooden sheath and a small central wheel made of some hard material such as lignite and mounted on an iron spindle. Originally, all blocks would have been made by hand, but as with rope making, the process was mechanised in the early nineteenth century. Marc Brunel broke down the process into individual steps, with a special machine for each operation – a very early example of mass production.

Not all barges were built in yards. The Constable painting (page 156) shows a simple dry dock cut out of the river bank and closed off by gates in which a Stour lighter is being constructed. Similarly,

Rope making in the eighteenth century. The man is teasing out the strands of hemp, while the boy turns the wheel that twists the fibres together.

when we turn to the canal age, we find that some boats were simply one offs, while others were built in well-equipped yards.

One boatyard has survived, having been working continuously since it opened for business at Banbury beside the Oxford Canal in 1778. Tooley's yard was made famous by L.T.C. Rolt when he described it in his acclaimed book *Narrow Boat*, published in 1944. He went there just before the start of the Second World War

BUILDING THE BOATS • 155

Launching a narrow boat. It was normal practice to launch them sideways.

to collect the narrow boat *Cressy*. 'Old Mr. Tooley' had started his working life as a boatman many years before he had taken over the yard. By this time, there was little demand for wooden boats but a good deal of business in repair and refurbishment. The site has a carpenters' shop, a dry dock and a forge. Rolt gave a fascinating insight into the craftsmanship and versatility of those who worked there. They had a generator to provide electricity to light the workshop and when a piston broke, they simply made a new one. This involved making a wooden pattern, a core and placing these in the mould box full of sand and then removing the pattern, which left the correctly shaped space to be filled with molten iron. The piston they made was every bit as good as the factory made one that had broken.

Construction of a wooden narrow boat is not dissimilar to that of any other wooden hull but made rather simpler by the fact that it has a flat bottom. Construction generally took place on a low, raised platform. First the bottom planks were laid down,

A painting by John Constable showing a Stour lighter being built in a simply constructed dry dock.

cut to size and in the correct order. Stem and stern posts were added. These would be shaped with an adze. The curved posts at stem and stern were generally steamed. These had a double curve, and Rolt was told that if both curves were produced by steaming, the result would be weak. Ideally the builder would find a piece of timber which already had one of these curves as part of its natural growth:

> Mr. Tooley must have carried this natural growth in his mind's eye, for he related how, years ago, he had spotted a suitable oak tree growing on the outskirts of the town, and when at last

he heard that it was about to be felled to make way for housing development, he bought it.

Once the boat was planked, the inside of the hold was lined with wood and the gaps between the two filled with chalico, an unsavoury mixture of tar, cow hair and horse dung, all heated together. Any iron parts that were required were made on the spot by the blacksmith at the forge.

The finishing touches for all narrow boats was the decoration, and at this point the whole family was involved, Mr Tooley and his two sons, George and Herbert. George began by painting the owner's name and port of origin on the cabin side, after which his father embellished it with garlands of roses. Finally, Herbert painted the traditional castles. 'Apart from striking a line with a chalked string to keep the lettering level, they did no preliminary sketching or spacing out whatever, but worked straight out of their heads with wonderful rapidity and skill.'

A narrow boat being repaired at Tooley's boatyard.

The nineteenth century saw many vessels that would once have been built out of wood being replaced by iron hulled vessels. For the next 100 years and more that would have meant riveting iron plates to an iron frame. Building an iron vessel, however, requires the sort of machinery that few of the old boatyards possessed. The start, however, is unchanged. Plans are drawn up and converted into wooden templates in the mould loft. This description describes the next stage of the process at Lairds shipyard on the Mersey at Birkenhead.

> These boards are in their turn taken down to the shop below, where the ship frames are bent, on perforated plates called levelling blocks, each measuring 10ft by 3ft and weighing several tons. In bending a bar of iron for a ship's frame, a small iron mould made from the lines on the large board, taken from the mould-loft, is placed on the blocks, and its outline drawn in chalk. The mould is then removed, and after pegs have been placed at intervals along each chalk line in the holes made for the purpose, the heated bar of iron is swung on to the block by crane power, and hammered to the required shape. The mould is then applied to the frame, which, when it is perfectly true, is removed to the punching press to have the rivet holes made in each arm.

Once the frame has been set in place on the ship, the job of riveting can begin. The rivet is an iron bolt with a rounded head. Holes in the plates are aligned with holes in the iron frame. The rivets are heated in a brazier to white heat and then thrown to one of the riveters. One man is on the outside, and he inserts the rivet and holds the rounded head firmly against the plate. On the far side, the second man beats the bolt until the end flattens out, firmly holding the plate to the frame.

For steamers, essentials such as the engine and boiler would be brought in from specialist companies, just as the diesel engines were supplied to narrow boats and barges. By the middle of the twentieth century, riveting was becoming obsolete and on today's narrow boats built for the leisure industry, hulls are of welded steel. Yet, through all these changes, it is good to know that one institution remains in business. After more than two hundred years, Tooley's is now run by a charitable Trust.

Chapter Eleven

The Modern World

Commercial traffic, apart from the holiday trade, has virtually ceased on the narrow canals of Britain and there is little chance of it ever returning. One reason is that so much trade now depends on containers – and even the smallest standard container is too large to fit a narrow boat. There have been attempts from time to time for individuals to use working boats to carry goods such as coal, which they sell around the system, but they scarcely count among the fleets of pleasure boats. There is, however, one major change in the pleasure boat world. Global warming has made everyone conscious of the need to get away from fossil fuels and move to more environmentally friendly forms of power. The result has been the appearance of electric narrow boats. These have been available for some time on hire from Castle Narrow Boats on the Brecon & Abergavenny Canal. Like many electric cars, they are plug ins. They have a range of around 18 miles but plug in points are available along the canal.

One of the pioneers of the electric narrow boat movement was Tim Knox, who was looking for an ecologically friendly boat and his first thought was to turn to the obvious choice – sail. But his wife was less than enthusiastic, so with the help of naval architect Andy Dovell, he designed an electric houseboat, but in this case not a plug in, but powered by solar panels. It was built in Australia and travelled the 90 kilometres of the Hawkesbury River using only solar power. This success led to him developing the Mothership Marine Company to design a boat with an array of ten solar panels built into the roof. The prototype, appropriately named *Shine*, made a highly successful 500 mile trip round Britain's waterways. The only problem with solar power is that Britain is not exactly a country with guaranteed sunshine, so a small generator is supplied to ensure the vessel always has power. They now offer a variety of craft for sale. This surely represents the future for holiday boating.

This may well be the future for pleasure craft on the waterways – a solar powered electric narrow boat.

The concept of using a tug to tow a succession of barges is not new, but in the latter part of the twentieth century, a new concept was developed known, somewhat paradoxically, as a push-tow. As the name half suggests, instead of pulling barges, the tug pushes them from behind. I first saw this operation on the Mississippi at Natchez, when a push-tow went past with the immense number of 40 barges in the train. This represents some 40,000 tonnes of cargo. In the UK, the maximum load for a single articulated truck is 41 tonnes; that push tow was carrying a load that on British roads would require a thousand trucks. Obviously, we do not have rivers the size of the Mississippi, but the lesson remains the same – efficient movement by water is better in many ways than movement by land with less road congestion and less fuel used. A modest start was made in Britain with push-tows, involving just three barges and a push tug, each barge carrying a maximum load of 140 tonnes. They might not be taking a thousand trucks off the road, but it was real contribution.

Barge traffic in Europe survives on a large scale, and in Britain there was an attempt to introduce the push-tow system on a modest scale. The most ambitious idea was to use a system that would allow barges to be moved from the heart of England to the centre of the continent. In the 1970s, a Danish company developed BACAT, short for barge aboard catamaran.

The catamaran was a twin hulled vessel. The push-tow would arrive with barges that would be floated between the hulls, some of which were lifted onto the deck, while the remainder were locked in place between the hulls. Once loaded, the BACAT ship would set off across the sea to its next port of call, where the barges could be floated out and continue an inland journey by river and canal in the new country. The intention was to use the system to allow, for example, barges to use the broad waterways of north east England.

It seemed an eminently sensible system, but there were problems from the start. The first was that at the time movement on the Humber was controlled by the British Transport Docks Board, while the inland waterways were run by British

In the mid-twentieth century, British Waterways introduced a new series of push-tow systems, where the tug worked three barges.

Waterways Board. The former charged the latter exorbitant fees, which threatened the viability of the whole system. Then, to make things worse, dockers at Hull and Goole could see their facilities being bypassed and decided to unofficially block the use of BACAT. There was no option other than to abandon the whole scheme. What could have been an interesting link with the heart of Europe was lost, though the BACAT barges continued to be used.

Among the many who were devoted to restoring barge traffic was Geoff Wheat, who died recently. I first met Geoff in the 1980s when he was owner of a Leeds & Liverpool short boat, *Lune*. Unfortunately, when being loaded there was an accident when the load was dropped and went straight through the hull. That did not deter Geoff from pursuing his enthusiasm for water transport, and he joined Humber Barges Ltd. They began operations on the Leeds & Liverpool in 1976 but would later switch to the waterways of the north east, with important contracts carrying aggregates to Leeds. In 2003 a new barge, *Fusedale H*, was brought into service. The craft, 175ft 6in long

The BACAT system. Here the tug is pushing two of the barges between the twin hulls of the mother ship, where it will be secured and taken to a port in mainland Europe where it will be released to continue its journey inland.

and 18ft 6in beam, was used in the aggregate trade and Geoff was the skipper. She still trades and a new agreement was made in 2020 to carry dredged aggregate from Hull to Leeds. The scheme is supported by Leeds City Council, whose executive spoke highly of the 'environmental benefits' of barge traffic. The benefits are there, but *Fusedale H* is the largest barge that can use the waterways – and it is unlikely that there will be any major changes to the system. The savings are impressive, but not as great as can be achieved in continental Europe. Looking at Belgium, for example, barge traffic is very significant and the trade had been showing a steady increase from 162 million tonnes a year to 205 million tonnes in 2018. New initiatives have been supported by the European Union – but, of course, we are no longer a member and cannot look for any help from that quarter. Will we see any similar increase in the use of our inland waterways? Will there be improvements to the system, and new initiatives, such as the use of solar power? It is difficult to be optimistic.

The Humber barge *Fusedale H*, one of the few still working on Britain's inland waterways, on the Aire & Calder Canal.

A hundred years ago, the inland waterways of Britain were vital links in the whole transport system of Britain. They were home to a rich variety of craft. Barges were still being worked under sail, even if most were also fitted with engines to cope with calms and to help with manoeuvring away from docks. The narrow boats and barges of the canal system were still being kept busy and even a very few horse-drawn boats could still find work. All that has gone, but we are lucky that so many craft have been preserved and run by enthusiasts. Sailing on a keel or a Thames barge may not be the experience it was for those who earned their living on those vessels, but they do at least give a glimpse into that way of life and, for many of us, instil a deep respect for those who worked them in their trading days. It is hoped that these historic craft will be preserved for posterity and that perhaps, in an age when energy saving will become paramount, there may yet be a way to bring our waterways back to commercial life.

Museums and Preserved Craft

MUSEUMS

Black Country Living Museum, Dudley
The museum is based on the Dudley Canal and is home to the steam narrow boat *President.* On site is a replica of a typical workshop created from an old narrow boat, turned on its side, and there are trips through the Dudley Canal tunnel. www.bclm.com

Canal Museum, Stoke Bruerne
The museum is sited next to locks on the Grand Union Canal. Inside are models of barges and narrow boats and the 1935 narrow boat *Sculptor* is moored outside. Boat trips are available locally. www.canalrivertrust.org.uk/places-to-visit/stoke-bruerne

Gloucester Docks and National Waterways Museum
The museum is housed in the former Llanthony Warehouse at the docks and has a wide range of canal exhibits, including archive films. Outside is the steam dredger *SND No.4* and the dumb Severn barge *Sabrina 5.* www.canalrivertrust.org.uk/places-to-visit/gloucester-docks

Ironbridge Gorge Museums, Ironbridge
The Blists Hill site includes part of the Shropshire tub boat canal and includes an incline plane and an example of an iron tub boat. It is also home to the last surviving Severn trow, *Spry.* www.ironbridge.org.uk

Linlithgow Canal Centre
The museum occupies former stables on the Edinburgh and Glasgow Union Canal and contains several models of local canal craft and also runs cruises on the canal. www.lucs.org.uk

London Canal Museum
The Museum is housed in a former ice house beside the Regent's Canal and exhibits include the back cabin of a narrow boat. www.canalmuseum.org.uk

National Maritime Museum, Greenwich
Although primarily concerned with sea-going craft, the museum also contains reconstructions of the excavations that revealed Britain's oldest river craft. www.rmg.co.uk/national-maritime-museum

National Waterways Museum, Ellesmere Port
Housed in the complex of warehouses at the transhipment centre between the Shropshire Union and Manchester Ship canals, the basin is home to the biggest collection of waterways craft in Britain, ranging in size from a Clyde Puffer to a starvationer that once brought coal out of the mines at Worsley Delph. Inside exhibits include *Friendship*, the former boat of Joe and Rosy Skinner, the last of the number ones.
www.canalrivertrust.org.uk/national-waterways-museum

River and Rowing Museum, Henley on Thames
The collection includes a variety of pleasure craft used on the Thames, including a traditional skiff. www.rrm.co.uk

Scottish Maritime Museum, Irvine
Much of the emphasis is on ship building, with original machinery from a Clyde shipyard. Among the preserved vessels is *Spartan*, the last Clyde Puffer to be built in Scotland. www.scottishmaritimemuseum.org

Tooley's Boatyard, Banbury
Not strictly speaking a museum as it is still a working site, the eighteenth century yard lies beside the Oxford Canal. www.tooleysboatyard.co.uk

Windermere Jetty Museum
The museum is home to forty boats, including an impressive array of steam launches and a very early motor boat built in 1898. www.lakelandarts.org.uk/windermere-jetty-museum

PRESERVED CRAFT

Humber Keel and Sloop
The keel *Comrade* and sloop *Amy Howson* are normally based at South Ferriby on the Humber. They are run by volunteers and trips and charters can be arranged through the preservation society. www.keelsandsloops.org.uk

Norfolk Wherry
The wherry *Albion* is available for charter from the Norfolk Wherry Trust for holidays on the Broads and there is always a competent skipper provided by the Trust. There are also occasional open days for the public. www.wherryalbion.com

Norfolk Wherry Yachts and pleasure Wherry
Hathor is a pleasure wherry built in 1905. Wherry Yacht Charter have restored this vessel, along with four wherry yachts. The yachts are available for charter, but the pleasure wherry is only used for educational trips and scheduled sailings. www.wherryyachtcharter.org

Thames Barges
There are many Thames barges preserved and a high percentage are available for short or long trips with competent skippers in charge. The Thames Sailing Barge Trust was formed to preserve these iconic craft and currently own two historic barges, *Pudge* and *Centaur,* both of which are available for short or long holiday trips. Other charter companies also operate and can be found online. www.bargetrust.org

Mersey Flat
Mossdale was built c.1860 and can be seen at the National Waterways Museum, Ellesmere Port (see above).

Tamar Barge
Shamrock 'built in 1899' is moored at Cotehele Quay, part of the National Trust Cotehele Estate. www.nationaltrust.org.uk/cotehele

Severn Trow
The last trow *Spry* is on display at the Ironbridge Gorge Museum (see above).

Canal Boats

The largest collection of canal boats of all kinds is to be found at the National Waterways Museum (see above). One exception is the steam narrow boat *President* at the Black Country Museum (see above).

Steamers

The paddle steamer *Kingswear Castle* is once again operating on the River Dart, where she first went into service in 1924. She is the last surviving coal fired passenger steamer in the world. www.kingswearcastle.org

Tattershall Castle is a paddle steamer that once operated a ferry service across the Humber. She is now a pub moored by the Thames Embankment in London. www.thetattershallcastle.co.uk

The Clyde Puffer VIC 32 offers five day holiday cruises for up to 12 passengers, usually starting from their base on the Crinan Canal at Lochgilphead. www.savethepuffer.co.uk

Select Bibliography

Carr, Richard GG, *Sailing Barges*, 1971
Chaplin, Tom, *The Narrow Boat Book*, 1978
Finch, Robert, *Sailing Craft of the British Isles*, 1976
Malster, Robert, *Wherries and Waterways*, 1978
McDonald, Dan, *The Clyde Puffer*, 1994
Paige, R.T., *The Tamar Valley at Work*, 1978
Schofield, Fred, *Humber Keels and Keelmen*, 1988
Smith, Peter L., *A Pictorial History of Canal Craft*, 1979
Tucker, Joan, *Ferries of the Lower Thames*, 2010
Tucker, Joan, *Ferries of the Upper Thames*, 2012

Acknowledgements

The author wishes to thank the following for permission to reproduce illustrations: A.D. Cameron. 118 top; Acabnasci, 120; Adrian Pingstone, 130(top); Andrew Gray, 118(bottom); Bodleian Library, Oxford, 140, 143; Cadbury Schweppes, 97; Clive Coote, 12, 13, 135, 152; Coventry Evening Telegraph,55; Dartmouth etc, 80, 81; Derek Pratt, 91; Gainsborough etc, 86; George Plunkett, 24; Gill Scott, 107; HM Queen,30; Humber Barges, 163; Humber Keel etc,16,17, 18; ICI, 92; James Potts, 96; Lancaster etc, 36; Library of Congress, 28; martinvi, 40, 151; NMM, 36; Michael Ware, 59; Mothership etc, 141; Newark, 15, 85; Nick Walker, 89; Phillip Lloyd, 52, 103; Robert Masters, 20, 147, 148; Rodw, 39; Stroud, 42; Thames etc, 38, 39; Tim Green, 130 (bottom); Transport Trust, 128; Trevor Hipperson, 25; Waterways Gloucester, 48, 56. 57. 60. 64, 70; Windermere, 137.

Index

Barge matches, 34-6
Boat builders and repairers: Press brothers, 26
Bridgewater, Duke of, 46-7
Brindley, James, 47-51, 62, 65

Canals: Birmingham, 50-1; Bridgewater, 46-9; Bude, 68-9; Caledonian, 66; Coventry, 50; Crinan, 66; du Midi, 47; Forth & Clyde, 65-6; Grand Union, 58-9; Huddersfield, 64; Kennet & Avon, 65; Ketley, 67; Lancaster, 64; Leeds & Liverpool, 62-4; Mersey & Irwell, 47; Oxford, 50, 52; Regent's, 53; Rochdale, 64; Sankey Brook, 6; Sheffield & South Yorkshire, 11, 14, 15, 16, 17; Shropshire, 67; Stainforth & Keadby, 14, 16; Staffs & Worcester, 50; Trent & Mersey, 50; Union, 65
Cartwright, Nell, 59, 62
Clayton, Thomas, 57-8
Craft: Bronze Age, 10; Humber keel, 10-18; Humber sloop, 18-19; ice breakers, 69-70; Iron Age, 9-10; lighters, 29-30; Mersey flats, 45, 48; Leeds & Liverpool short boats, 63-4; maintenance boats, 69; narrow boats, horse drawn, 50-62; narrow boats, iron, 69; Roman, 10-11; scows, 66; starvationers, 48; Stour barges, 42-5; tar boats, 57-8; Tamar barge, 38-40; Thames barges, 30-7; trows, 39-42, tub boats, 67-9; wherries, cargo, 19-26 wherries, passenger and pleasure, 19, 26-7
Craft, named: Albion, 20-6, 28, 34; Amy Howson, 18-19; Centaur, 39; Comrade, 10-18; Emma, 31; Ena, 36; Pudge, 36-8; Shamrock, 38-40; Spry, 42; The Trial, 69

Dodd, Henry, 34
Dunkirk, 36-7

Ellesmere Port, 52

Fellows, Morton & Clayton, 58
Flash locks, 11, 31

Gilbert, John, 47

Harecastle tunnel, 50
Henrik, Frederick, 33
Humber Keel & Sloop Preservation Society, 17

Inclined planes, 67-9

Keadby, 16

London bridge, 29
Longbottom, John, 62-3

Mikron Theatre Company, 62
Mother Shipton, 69

Norfolk Broads, 19-28
North Ferriby, 9

Pickford's, 53
Powley, Walter, 25

Reynolds, William, 66
Rivers: Avon, 8 Humber, 9, 18; Irwell, 47; Lee, 11; Medway, 33; Mersey, 45, 48; Tamar, 38-9; Severn, 39-42; Stour, 42-5; Thames, 29-36; Weaver, 45
Runcorn, 48

Schofield, Fred, 11, 13-14, 17-18
Sheffield, 16
Skinner, Joe and Rose, 52
Smeaton, John, 65
Stonehenge, 8

Taylor, John, 30-1
Thames Navigation Committee, 31

Wilkinson, John, 69

Yarmouth, 19, 20, 25